GLACIAL AND FLUVIAL GEOMORPHOLOGY OF WESTERN HIMALAYA

(Liddar Valley)

GLACIAL AND FLUVIAL GEOMORPHOLOGY OF WESTERN HIMALAYA

(LIDDAR VALLEY)

MANMOHAN NATH KAUL

CONCEPT PUBLISHING COMPANY, NEW DELHI-110059

ISBN 81-7022-244-3

First Published 1990

Published by
Ashok Kumar Mittal
Concept Publishing Company
A/15-16, Commercial Block, Mohan Garden
NEW DELHI - 110059 (India)

Lasertypeset by
Microtech Advance Printing Systems (Pvt.) Ltd.
H-13, Bali Nagar
NEW DELHI - 110015

Printed in India by
Taj Press
DELHI - 110006

Foreword

More than 200 years of observations and research on the region notwithstanding the Himalaya continue to be, for most parts, a classic *terra incognita.* It has yielded to persistent probings during the late 19th and early 20th centuries but only slowly and to a very little extent. The forbidding diversity and complexity of physiography and culture have lured into the Himalaya only the intrepid explorers and scientists. For others the land still borders on the mysterious. The layers upon layers of ranges have spawned and preserved in them myriads of fictions. We know little of the facts. While we know little of the facts our knowledge of the elements of physiography is still more limited, far too limited than our knowledge of the people and culture. The enormous corpus of literature existing on the Himalaya is more of an inchoate mass of description ranging from cabbages to the kings.

When considering the physiography of the Himalayas the element that stands out most prominently is the complex, polygenetic geomorphology. As it is the Indian geographers and geologists have shunned, by and large, the investigations in geomorphic landscapes, their character and their evolution. The Indian geographers have concentrated heavily on the extra-Himalayan areas for the obvious reason of being far more easily accessible than the Himalayas–forever uncertain and seemingly treacherous! The geologists have for long been interested and, to a great extent, continue to be so in the pre-Quaternary periods. Even tertiary has secondary interests for them. The Himalayan geographic landscape is of little attraction for them because it is almost entirely the product of Quaternary processes and conditions, the most important among them having been those of the glaciers, rivers, and masswasting. Although these are significant agents their distributional patterns are markedly different. The glaciers are located up in the sheltered valleys of the high altitude ranges. The glaciers have thus been more inaccessible and the last to receive scientific consideration. It is in two perspectives that Dr. M.N. Kaul's book should deserve deep critical notice and appreciation : (i) this is the first book-length study of the glacial geomorphology of a large glaciated tract in the Himalaya, and (ii) this is the only comprehensive study of glacial geomorphology of Himalaya that is based almost entirely on field observation, experimental data, and measurements.

The Himalayan glaciers have been investigated over scattered areas in a perfunctory manner with a major focus on the individual geomorphic features related directly or indirectly to the processes and phases of glaciation. Different aspects have indeed been described and interpretations attempted by different scholars in their published papers but there has been no comprehensive study of a single glacier system. Dr. Kaul's subject of research is the Liddar Valley glacier group.

That the Liddar glaciers, numbering 17, have completely escaped the notice of scholars (particularly the geographers) is amply borne out by the fact that a comprehensive compendium such as undoubtedly Spate's *India and Pakistan* does not take any cognizance of their presence. Nor are these glaciers small, moving ice blocks; they have created long glaciated valleys, moraines, drumlines, and *roche moutonnees* in abundance and in whole fields which Dr. Kaul has so meticulously observed, identified, and documented. The Liddar Valley abounds in aretes, cirques, and horns, all created directly or indirectly by glaciers and the processes generated by glaciers. Occurring in the same general terrain can be observed fluvial features dominated by fluvial depositional terraces. Gorges and glacio-fluvial features are equally prominent.

Dr. Kaul does well by raising some questions and stating clearly his objectives but not embroiling us with a large number of hypotheses. In this he follows the methodological structure displayed in the more significant of the geomorphological research. In geomorphology there are questions to be answered and an individual question may arouse a set of multiple hypotheses but the entire, long thesis really is not founded upon a few hypotheses but a series of questions. Dr. Kaul nowhere states the questions explicity but these are clearly embedded in the objectives and as he plods to realise his objectives through comprehensive investigations and clearly identifiable steps the questions begin to dawn on a scholarly scrutiny. By avoiding the confounding jugglery of methodology and adopting consistently empirical, inductive fieldwork and analysis Dr. Kaul has made an outstanding contribution to the meagre corpus of glacial geomorphology of the Himalaya in particular and that of the geomorphology of the Himalaya in general. Nor is he writing for the novitiate, the uninitiated, for he uses sophisticated field methods and sensitive measurements which can be evaluated by scientists familiar with field techniques of glaciology. In following the footsteps treaded by the great masters in geomorphology he spares the prospective scrutiniser of the desperation induced by the mind-boggling rabble rousings of quantification. Dr. Kaul's book is based on the primitives of both geography and geomorphology: What, where, how much, how, and why are the questions around which he organises the investigations and presentations. These are the elements of inductive not deductive geomorphology for Dr. Kaul does nowhere go beyond the limits of observation, measurement, and analysis-based interpretation into the realms of conjecture. On the way he demolishes many old faiths and collects many new facts. He is not transferring the knowledge of other glaciers he could acquire from secondary sources to the glaciers of the Liddar Valley. His comparisons are limited to the Himalayan glaciers and his glaciers are limited to the Liddar Valley.

Dr. Kaul's book would merit serious consideration not only because this is the first publication in the field of glacial geomorphology produced by an Indian geomorphologist or geographer, though that itself is a significant achievement, but more so because he investigates some of the most important fundamental problems of this specialisation which include mass-balance of glaciers, history of glacial dynamics, and palaeogeographic reconstructions in which he has used, with great effects and impressive results, some of the latest, state-of-the-art techniques including lichenometric and depth-of-snow sounding, fossil flora, and pollen analysis. Let us look at the book in brief detail to support the claim made by me in the preceding line.

Dr. Kaul begins his book with a detailed and neat description and statement of the objectives and the essential geological elements which will be subsequently correlated with the different aspects of glacial and fluvial geomorphology of the Liddar Valley but he reminds us that the necessary geological discussion must be organised in only three basic components: lithology, stratigraphy and structure, if they are to be of maximum use to geomorphology. Along with this important innovation Dr. Kaul also introduces the new mode of describing lithology by using the dominant rocks and rock groups in place of the conventional geological formations widely used by the geologists. There is a separate but brief discussion of the lithology and stratigraphy of the selected sites of Liddar Valley, in itself of great relevance to geomorphology. Finally, Dr. Kaul arranges the material of the three components of geology in an inter-related manner. The structure and stratigraphy of each geological formation has been dispensed with an instead the structural details that are reflected in the geomorphological features and characterised are discussed in necessary details. Thus, the structure and lithology are both interpreted in terms of geomorphology.

Dr. Kaul very logically begins with a chapter length discussion of the glacial dynamics of the Liddar glaciers. Discussing at length the distribution of each glacier in terms of setting Dr. Kaul already brings home the more fluential factors in the advance and retreat and snow gain through accumulation and snow loss through ablation. For each of the 17 glaciers of the Liddar Valley Dr. Kaul gives us the primary data of attributes collected by him from his field measurements and computation from large scale toposheets, aerial photographs and largescale resolution of satellite imageries: area, direction of snout from the catchment area, length, breadth near snout, extension of glacier, height of snout, height of source area, and relative relief. It is remarkable that Dr. Kaul could muster up courage for conducting field investigation and measurement during winter to record the advance and retreat through seasons for each and every of the 17 glaciers. His use of stakes in measuring the movements is starkly simple and must be followed by every scientist of glaciers. Dr. Kaul introduces a major innovation in this chapter; for at least 9 glaciers (4 of the West Liddar Glacier Group and 5 of the East Liddar Glacier Group) he has constructed horizontal and vertical zonation by erecting an attenuated triangle in a matrix of area (in square kilometre) on 'X' axis and elevation (metres) on 'Y' axis. In this matrix (one triangle for a glacier) he charts as many as eight attributes: area, height of glacier snout, orientation of glacier, maximum glacial advance, tree line, winter snowline, summer snowline and glacier extent. Easily this diagram is one of the most innovative and of immediate use in correlation statements and comparative analysis of the West and East Liddar glaciers.

In this section the most impressive and important contribution is the innovative discussion of mass balance of the Liddar Valley glaciers. Not many geologists, geomorphologists and glaciologists had earlier paid attention to this very important aspect of glacial dynamics in India. Equally commendable is the formulation of a new model (shown in a clearly drawn and reproduced sketch) of the system of mass balance. The property or phenomenon of mass balance is particularly suited to being treated as a system — a spatial system as well — in which the two main sub-systems have been identified: glacier storage and glacier processes — convincingly demonstrated are the positive and negative feedback processes which sustain or prevent the survival and continuity of the mass-balance system. The two zones of ablation and accumulation having different sites automatically lead to the emergence of a spatial system. In order to compute mass-balance Dr. Kaul uses the sounding method — rarely used by his Indian colleagues — to determine the depth of accumulated snow, an important element of the mass-balance calculation. From the analysis of snow samples dug out from different depths Dr. Kaul determines as many as six elements length of snow sampler, length of sample snow column, weight of the snow sample column, density of snow sample, water equivalent of snow sample and cumulative water equivalent of snow pit samples. One would be impressed with the high degree of scientificity in these field exercises with unmistakable clarity and precision. He repeats the field exercise for both Kolahoi and Shishram glaciers by digging 4 and 3 pits respectively and from each pit he takes as many as 5 to 11 samples. In order to collapse the large data body into easy to comprehend diagram Dr. Kaul uses 'depth-density relationships in snow pits' graphs remarkable for their simplicity and power of communicating a none too easy relationship.

Then he proceeds further and comments on the entire set of data in two maps showing by choropleth shades the winter accumulation in water equivalent in the two glaciers, Kolahoi and Shishram. For measuring ablation Dr. Kaul uses a simple method of stakes fixed through the snow. There is such an abundance of innovative diagrams and sketches and maps as to defeat the efforts of a reviewer to select the most prominent one. The diagram (graph) displaying the net balance derived from winter balance and summer balance immediately reminds one of the famous Thornthwaite water-balance diagram but put to an altogether different and unexpectedly new purpose. To further substantiate his findings on the mass-balance of the two major glaciers (Kolahoi and Shishram) Dr. Kaul takes up a rather neat and compact analysis of the hydrological model, thus, also establishing the significance of this analysis to the river regime (discharge), which in itself has a larger dimension of social relevance, thus bringing physical and human geography into an interactive system. Also Dr. Kaul proceeds promptly from this point to the preparation of a map of physiographic divisions that shows three zones : glacier, ablation and rain. Equally fascinating is the documentation presented by Dr. Kaul on the retreat of the glaciers. For this he calls out evidences from a large number of sources : glacial geomorphological, sedimentological and historical. He uses an innovative method with great effect to reconstruct the advance and retreat of the major and well-known glaciers, Kolahoi and Shishram. This method involves the comparing of photographs of contemporary (earlier) stadial positions with those of the modern on the one hand and the paintings done by landscape painters of the nineteenth century with the photographs of the modern times. Both the comparisons strikingly reveal the advance but more particularly the retreat of the snout. What makes the method at once scientific and unassailable is that there is a complete convergence of the evidences on to a common conclusion.

The dynamics of the Liddar glaciers discussed in superb and controlled detail, Dr. Kaul moves down to the current landscape and for that he introduces a new organisation of contents: he devotes three chapters to (i) glacial erosion, processes and their geomorphological features, (ii) glacial transportation, processes and their geomorphological features and (iii) glacial deposition, processes and their geomorphological features. Instead of beginning with a long cumbersome and unrelated inventory of all the features without grouping them as to their origins and then describing each of them to produce an encyclopaedic body of information Dr. Kaul has organised the features in terms of their origin in three separate chapters but the selection of the features is logically on the basis of their significance in the understanding of one of the three major glacial processes: erosion, transportation and deposition. Two points of major significance arise from such an organisation: the features produced by glacial transportation are isolated and specially discussed in sharp contrast to the prevailing corpus of analysis in which glacial transportation is consistently ignored in the formation of features and the geomorphological interpretation of features produced by any other sub-aerial, exogenous agent can be organised into three clusters based on erosion, transportation, deposition. At least a search for the features produced by the transportation processes of an agent should become necessary. Dr. Kaul finds that the glacial landscape of the Liddar Valley has developed on a pre-glacial topography that was carved out by fluvial processes. In this Dr. Kaul answers emphatically to the latest mode of explanation in glacial geomorphology. This is further strengthened by establishing that the Liddar Valley geomorphic landscape is a polygenetic form, a discovery of great importance to the polyclinic-polygenetic debate. With the help of an original model and sophisticated mathematical mode Dr. Kaul explains the nature of glacial erosion, its effectiveness and the resultant features produced at different altitudes. The model of glacial erosion is not merely heuristic, it can be immediately translated into quantitative terms for all the variables from quantitative entities and can be measured, either directly by instruments or indirectly by surrogate — involving formulae. As a matter of fact the model of erosion and erosional landforms is one of the most original contributions of Dr. Kaul to the entire gamut of glacial geomorphology and surely does proud the Indian geomorphologist colleagues. The model is a representation of a system with very clearly distinguished sub-systems separated one from the other by clear boundaries so that the results of one sub-system passes on the other but located at a lower hierarchical layer: the whole system thus forming a convincing example of a cascading system. It is a hierarchical structure of inter-related determinants. As a true practitioner of genetic geomorphology Dr. Kaul proceeds from the processes to features and yet nowhere does he deviate substantially from his chosen inductive approach in which the attributes of processes form the foundation of the structure of analysis. Throughout the three chapters he introduces highly original mathematical graphic representations of the features, their orientation, sites of rockfall and distribution of glacial features. As a matter of fact there is little in the text that Dr. Kaul has not converted into maps, tables, sketches and photographs, the former three being consistently original.

Since Dr. Kaul has already established the polygenetic nature of the glacial landscape of the Liddar Valley he devotes two more chapters to glacio-fluvial landscape and fluvial landscape. He separates the zone of melt-water channels from that of the fluvial channels, not quite obvious to an observer of the scene. Here again Dr. Kaul introduces explanatory sketches and diagrams to describe and analyse both the properties and the origin of the features and in reaching this stage he casts his net very wide to collect his evidences. He undertakes a detailed reconstruction of soil profiles in order to convincingly portray the evolution of eskers, kame terraces and kame delta. Terraces belonging to three levels and gorges are his major discoveries. Morphostratigraphy of the terraces and the river-capture origin of the gorges are his innovative contributions.

Easily one of the most important contributions of Dr. Kaul in this book is the palaeogeographic reconstruction of the Liddar Valley glacial landscape. To do this he uses the results of the analysis he has himself conducted on fossil flora, pollen analysis, lichenometric analysis and evidences of glacial geomorphology. Mainly through these evidences, Dr. Kaul, the first time ever done, charts out the history of climatic and vegetational changes, the evolution of the landscape and the changes in the related geomorphic processes. From these historical reconstructions Dr. Kaul succeeds superbly in his mapping of the glacial extents and stadial positions at different periods.

Throughout the book Dr. Kaul retains consistently a genetic approach emphasising the sequential evolutionary mode of analysis. This immediately relates the processes, the features and the historical stages and the climatic conditions. Although genetic he steadfastly holds on to an inductive approach in which he moves from the processes to the features. In essence the method is correlative and inter-relative. Dr. Kaul uses highly innovative observational, experimental and measurement techniques and quantitative methods ranging from the simplest to the most sophisticated currently available. Some of the methods have been used for the first time by an Indian geomorphologist in the Himalayan context. Throughout the book one is deeply impressed by the super support given by the maps, charts, aerial photographs, terrestrial photographs, tables and graphs. Complex relationships have been expressed in simple representations.

There is no question that Dr. Kaul's book is the first full-length treatment of the glacial geomorphology of a fairly large valley of the Himalaya. Thus, it serves as a model to other scholars who are invited to cultivate a new field where they will not be strangulated by the sterile Hortonian and Strahlerian mode of quantitative analysis. Dr. Kaul's study will prove to be of immense value to glaciologists, glacial geomorphologists, landsat image and remote sensing specialists, hydrologists, flood-management experts and environmentalists.

A.B. MUKERJI

Contents

List of Figures

List of Tables

Acknowledgments

The book has grown out of my Ph.D. thesis which was prepared under the supervision of Professor Dr. A.B. Mukerji, Department of Geography, Panjab University, Chandigarh (India). I express profound gratitude to Professor Mukerji for suggesting this problem and providing patient guidance, constant encouragement and careful reading of the manuscripts. This book is dedicated to him.

My thanks are also due to many officials of the Power Development Department, Srinagar; Geology and Mining Department, Srinagar; and Survey of India, Chandigarh, for providing necessary documents and data.

I wish to thank the University Grants Commission, New Delhi, for supporting me with a Teacher Fellowship during the tenure of the research on which the book is based.

I am grateful to my parents, wife Santosh and daughter Deepa, for sustaining me with unstinted encouragement and moral support in the accomplishment of my aim.

CHANDIGARH
March, 1989

MANMOHAN NATH KAUL

1

Introduction

The Himalayan region, endowed with a staggering geomorphological diversity, offers an exceptionally rich field for the study of glacial and fluvial landscapes. Its active glaciers and associated landscapes provide an ideal laboratory for the investigation of fluvial and glacial processes and their resultant land forms. The study of the current processes throws light on the evolution of land forms created by similar processes in the past.

There are nearly 15000 glaciers in the Himalaya, extending from Kashmir in the west to Sikkim in the east. The western Himalayan glaciers contribute nearly 70 per cent of the annual flow of the major rivers of northern India in the form of meltwater and the Eastern Himalayan glaciers contribute relatively less due to high snowline and intense monsoons.[1] On an average a Himalayan glacier produces 0.1 cubic kilometre of water during the summer season. This is of great economic value, as the melting reaches its maximum in the season when the non-perennial (seasonal) rivers dry up.[2] There has been a general retreat of glaciers in the Himalaya since the beginning of the present century. This is analogous to the global retreat of glaciers due to an increase in warmth and decrease in snowfall in the catchment areas. Due to these changes there has been a net reduction in the thickness and length of glaciers which have ultimately affected the rate of meltwater production and geometry of the channels.

STUDY AREA

The present study deals with the fluvial and glacial geomorphological investigations of the Liddar (also spelled as Lidder in the early toposheets of the Survey of India) valley of the Kashmir Himalaya. The Kashmir Himalaya forms the westernmost extremity of the main Himalayan range. The Liddar Valley occupies the southeastern part of the giant Kashmir Himalayan synclinorium and forms part of the middle Himalaya.

Locational Setting

The study area lies between the Pir Panjal range in the south and southeast, the North Kashmir range in the main Himalayan range in the northeast and Zaskar range in the northwest (Fig. 1). The area gradually rises in elevation from south (1600 metres) to north (5400 metres). The Liddar valley is girdled on three sides by lofty ridges, the

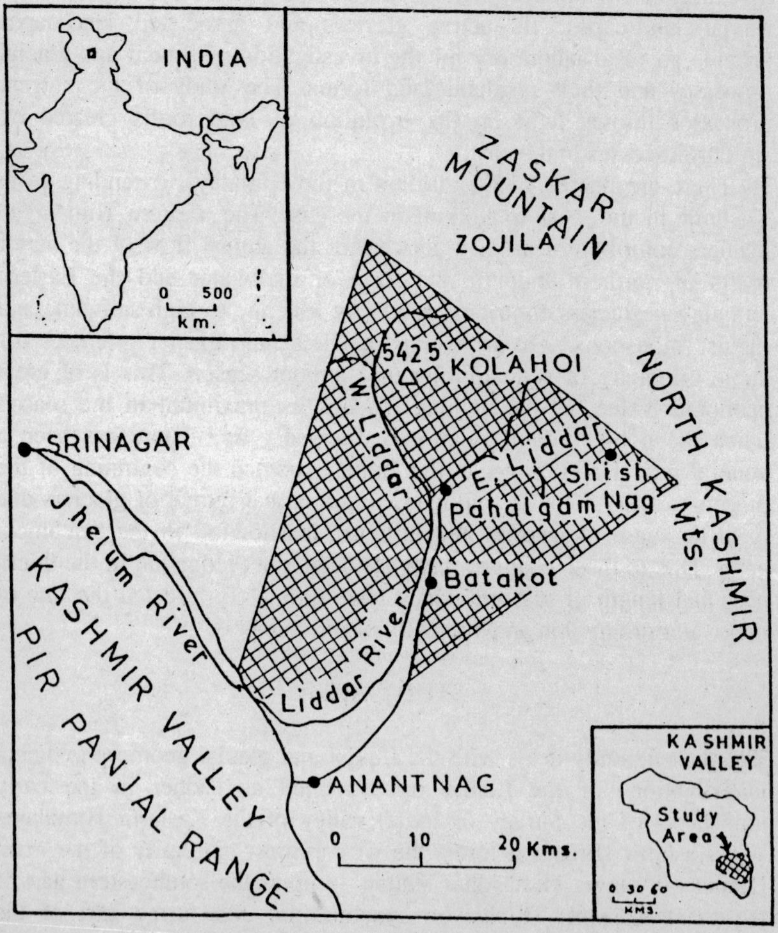

Fig. 1: *Location of Liddar Valley*

Saribal-Katsal ridge (4800 metres) on the east, the Wokhbal (4200 metres) on the west and Basmai-Kazimpathlin bal (4800 metres) in the north. The interior of the northern part of the Liddar valley has a concentration of high mountain ridges like Dadwar (3560 metres), Goucher (4000 metres) and Tramakzan (4200 metres), which run parallel to the course of the rivers (Figs. 2 and 3).

The Liddar valley begins from the base of the two main snow fields, the Kolahoi and the Shishram. From here its two main upper streams, the West and the East Liddar, originate and join near Pahalgam. Below Pahalgam the Liddar passes through a narrow valley and ultimately debouches on to the wide alluvial fan near Aishmaqam. The Liddar meanders and forms a braided valley in the plains (Fig. 4). The Liddar valley reveals a variegated topography due to the combined action of glaciers and rivers. At present the glacial activity is mostly confined near the snouts of the Kolahoi and Shishram where active glaciers are located. The glaciated section of the valley contains many erosional and depositional features. The remnants of these features extend parallel to the Liddar river. In the South Liddar valley the river terraces are prominently developed in some sections.

Climate

The Liddar valley has distinct climatic characteristics. This is related to its location at higher altitudes and its setting which is enclosed on all sides by mountain ranges. The study region has sub-mediterranean type of climate with nearly 80 per cent of its annual rainfall concentrated in winter and spring months.[3] The study area has CZW moisture regime (Thornthwaite) and the climate type designated as B′1 with thermal efficiency index of 55 per cent.[4]

The rainfall and temperature data of the study region are available only for four raingauge stations which have been installed recently. The only observatory at Srinagar (70 kilometres from Pahalgam) remains open throughout the year. The detailed analysis of the main weather elements (temperature and precipitation) for the period 1973–80 shows that high mountain terrain exerts an important influence on local weather processes. The temperature and precipitation change after every hundred metres of altitude. There are large seasonal variations in the mean maximum and mean minimum temperatures and precipitation.

There is a consistency in the variations in monthly mean maximum temperatures at different stations. This is related to the variations in altitude. The variations in the minimum temperature are very large within the high mountain areas.[5] This variation is due to the local topography of the site. The temperature variations suggest marked seasonality in the weather, with a long cold season lasting for nearly 7 months and a short summer season (Table 1A).

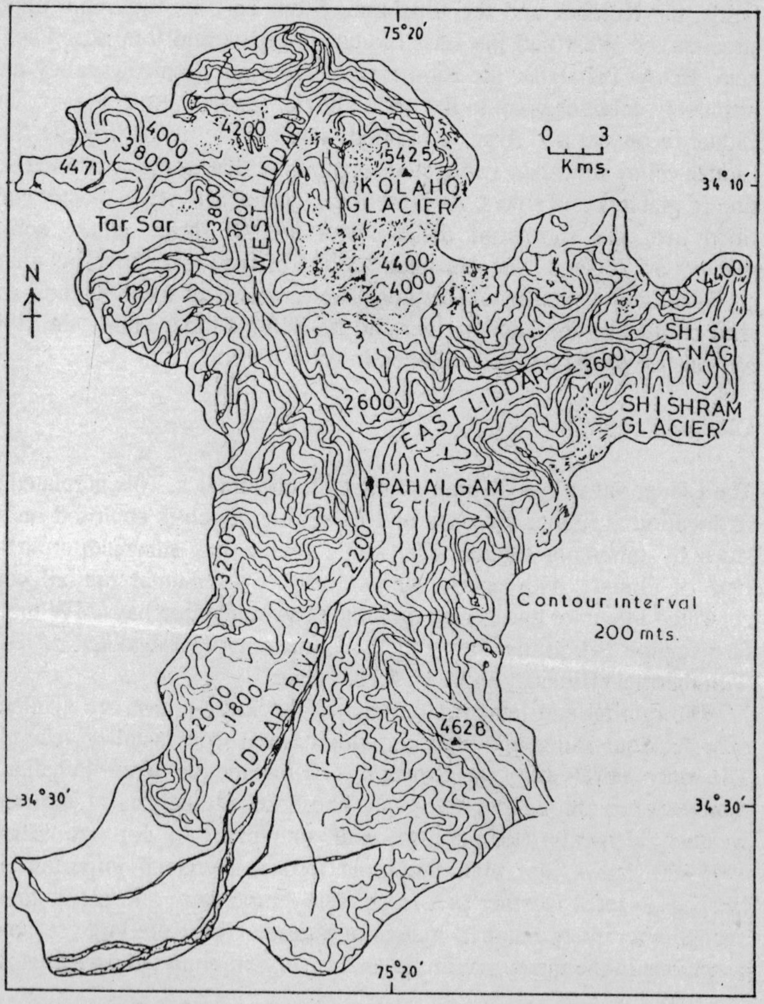

Fig. 2: *Liddar Valley Contour Patterns*

Fig. 3: *Satellite imagery of Liddar Valley shown in Dotted Lines. It is bounded in north and east by two ice fences (A), glacial landscape (B) and fluvial landscape (C)*

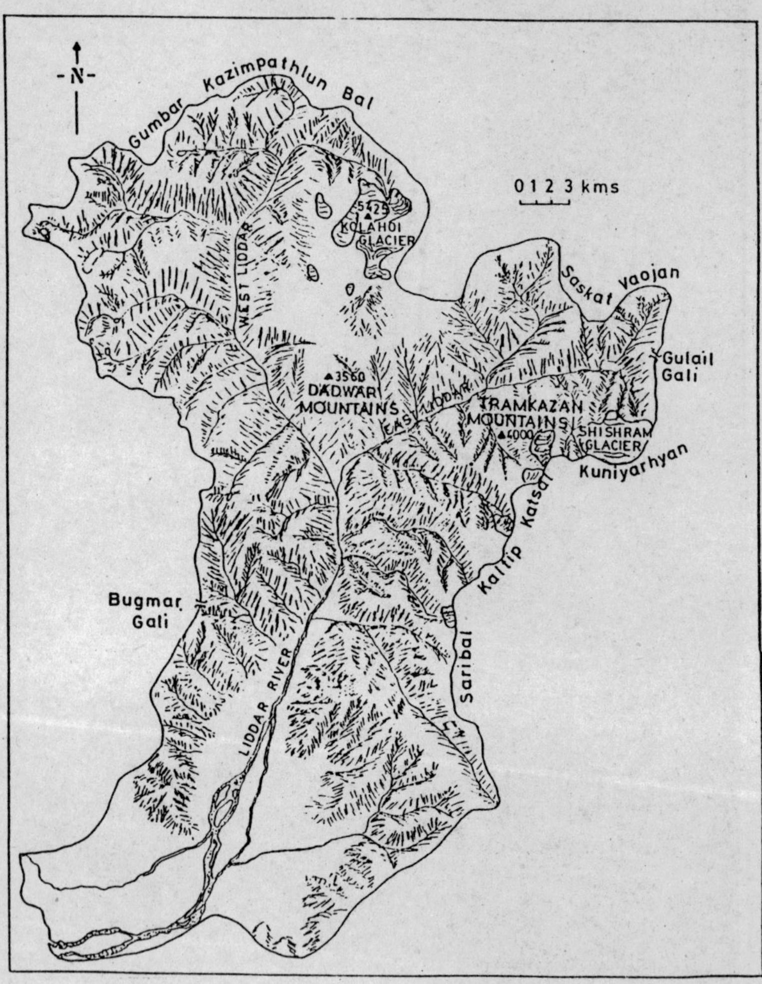

Fig. 4: *Liddar Valley: Physiographic Setting*

Table 1A: *Mean Monthly Temperature and Rainfall of the Selected Sites of Northern Part of the Liddar Valley: 1973—80*

Months	*Pahalgam (2200 metres, above sea level)*			*Pharsulan (2300 metres, above sea level)*			*Chandanwari (2700 metres, above sea level)*			*Aru (2400 metres, above sea level)*		
	Temperature (°C)		*Rainfall (mm)*	*Temperature (°C)*		*Rainfall (mm)*	*Temperature (°C)*		*Rainfall (mm)*	*Temperature (°C)*		*Rainfall (mm)*
	Max.	*Min.*		*Max.*	*Min.*		*Max.*	*Min.*		*Max.*	*Min.*	
January	5.1	–8.6	122	4.2	–10.9	135	1.9	–12.3	172	3.1	–11.7	146
February	6.8	–9.1	72	5.6	–11.3	82	3.1	–13.2	84	4.2	–12.6	92
March	14.7	–2.3	124	12.9	–3.5	136	11.9	–4.7	153	12.2	–4.0	126
April	20.0	2.9	96	18.8	2.0	105	16.8	1.1	122	17.7	1.7	100
May	24.8	4.9	78	23.9	3.8	94	20.9	3.2	105	25.1	3.6	98
June	30.0	5.1	142	29.0	4.2	167	24.6	3.9	194	26.4	4.7	170
July	29.5	10.7	68	26.3	8.2	82	23.4	5.8	99	25.2	6.8	88
August	28.3	10.5	64	24.7	7.6	73	22.2	5.6	86	23.8	6.1	74
September	26.2	8.2	16	22.3	5.7	18	20.2	4.9	34	21.4	5.2	17
October	22.3	0.9	11	17.1	–1.4	14	15.4	–4	16	16.9	–3.2	15
November	14.1	–4.9	4	12.4	–5.8	6	9.8	8.6	8	11.7	–7.1	6
December	6.4	–6.7	88	4.9	–8.1	115	2.2	–11.2	164	4.5	–10.4	122

Source: Indian Meteorological Department, Srinagar, Monthly Weather Reports.

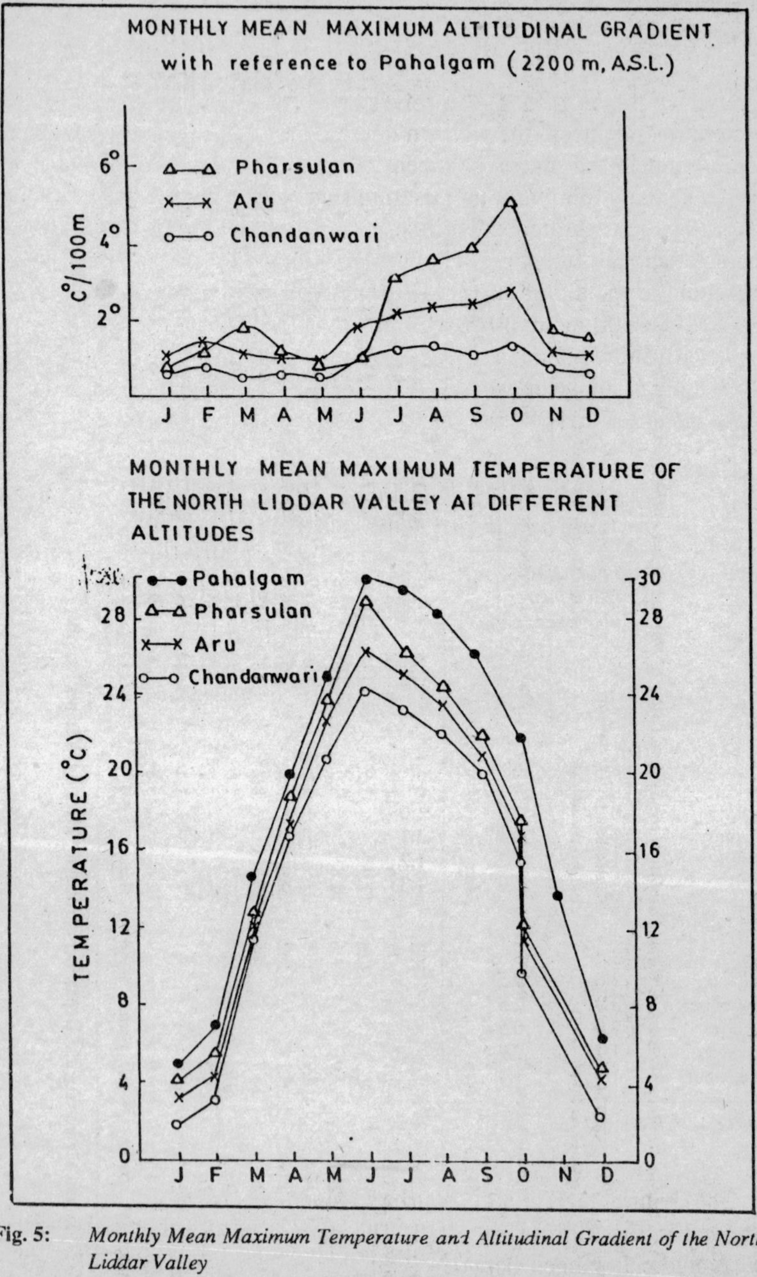

Fig. 5: *Monthly Mean Maximum Temperature and Altitudinal Gradient of the North Liddar Valley*

Winter Season

The winter season is long. It receives heavy precipitation caused by the western disturbances (the western depressions). The normal daily maximum temperature ranges between 2°C and 20°C during November to May. The daily minimum temperature fluctuates between – 13.2°C and +1°C. The temperature distribution in the higher altitudes shows characteristic altitudinal gradient. The variability of monthly mean maximum of the altitudinal gradient is large (standard deviation of 0.9° to 1.8°C per 100 metres) between Pahalgam (2200 metres above mean sea level) and Pharsulan (2300 metres above mean sea level), Aru (2400 metres above mean sea level) and Chandanwari (2700 metres above mean sea level) respectively (Table 1B) (Fig. 5A).

Table 1B: *Mean Monthly Altitudinal Gradient of Temperature of Selected Sites of the Northern Part of the Liddar Valley with Reference to Pahalgam (2200 metres, above sea level): 1973-80*

Month	*Pharsulan (2300 metres, above sea level)*		*Aru (2400 metres, above sea level)*		*Chandanwari (2700 metres, above sea level)*	
	Temperature Altitudinal Gradient (°C/100 m)		*Temperature Altitudinal Gradient (°C/100 m)*		*Temperature Altitudinal Gradient (°C/100 m)*	
	Maximum	*Minimum*	*Maximum*	*Minimum*	*Maximum*	*Minimum*
January	0.9	2.3	1.0	1.2	0.6	0.7
February	1.2	2.1	1.1	1.8	0.7	0.8
March	1.8	1.2	1.2	0.8	0.5	1.0
April	1.2	0.9	1.3	0.6	0.6	0.5
May	0.9	1.1	1.0	0.7	0.6	0.4
June	1.0	0.9	1.8	0.2	1.1	0.3
July	3.2	2.5	2.2	1.9	1.2	0.4
August	3.6	3.5	2.3	2.2	1.3	1.0
September	3.9	2.5	2.4	1.5	1.2	1.0
October	3.9	0.5	2.7	2.2	1.4	1.1
November	1.7	0.9	1.2	1.3	0.8	0.7
December	1.5	1.4	1.0	1.9	0.4	0.9

Source: Calculated from Monthly Weather Reports, Indian Meteorological Department, Srinagar.

The winter precipitation during November to March is mostly received in the form of snowfall. The pattern of snowfall varies with the altitude. It ranges between 1 to 4 metres from Pahalgam (2200 metres above mean sea level) to Kolahoi-Shishram areas (3800 metres above mean sea level) respectively, as recorded in the field during the 1983-85 field season.

The monthly mean minimum of the altitudinal gradient ranges between 1° and 3.5°C per 100 metres (Fig. 6A). It is minimum during the winter months of December to February. The winter months experience larger precipitation. *Chandanwari, Aru, Pharsulan* and *Pahalgam* received *820, 730 and 580* millimetres of rainfall from November to May respectively.

A study of the temperature records from 1973 to 1980 reveals steep fall in the night temperature during winter season followed by cold wave. The night temperature commonly deviates by 10°C to 20°C from the normal during November to March.

Summer Season

The summer season commences from June and terminates in October. The temperature curves show that the mean monthly maximum ranges between 16.5°C and 30°C. The period from June to September records the highest temperature range between 26°C to 30°C at Pahalgam and 21.2°C to 25.6°C at Chandanwari (Fig. 5). The variability of monthly mean maximum temperature of altitudinal gradient ranges between 1°C and 3.5°C per 100 metres. It is maximum during August to October and minimum during June and July (Fig. 5).

The monthly mean minimum temperature of the altitudinal gradient ranges between 1°C and 2°C per 100 metres (Fig. 6).

The summer and winter rainfall variations with altitude are similar. It is the highest (194 millimetres) at the higher altitude of Chandanwari and the lowest (142 millimetres) at the lower altitude of Pahalgam. The rainfall is local in nature and gets marginal benefit from monsoonal winds because of the rain-shadow effect of the Pir Panjal.

Vegetation

The vegetation of the study region displays characteristic mixture of temperate and alpine flora. The phytogeographic composition of plant communities is different at different elevations due to microclimatic, biotic and other environmental conditions. The following vegetation zones have been identified by botanists working on the Himalayan areas[6,7] (Fig. 7).

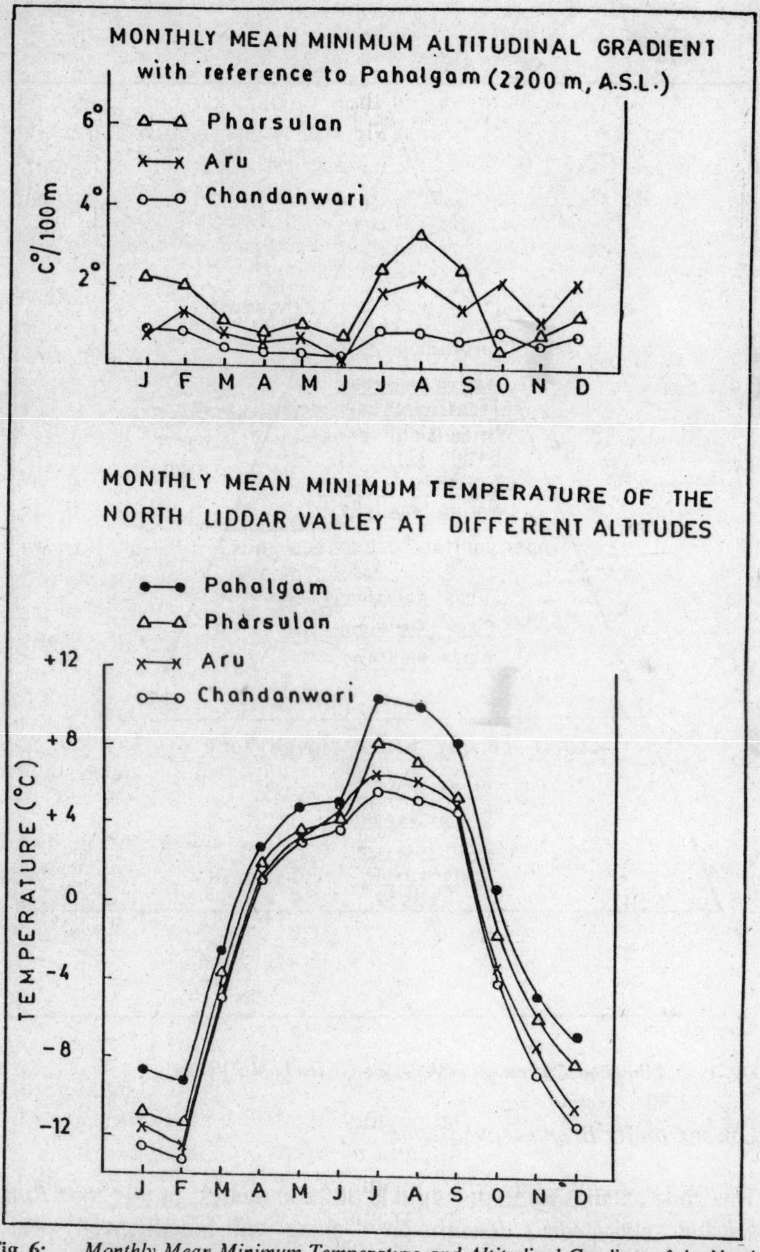

Fig. 6: *Monthly Mean Minimum Temperature and Altitudinal Gradient of the North Liddar Valley (below)*

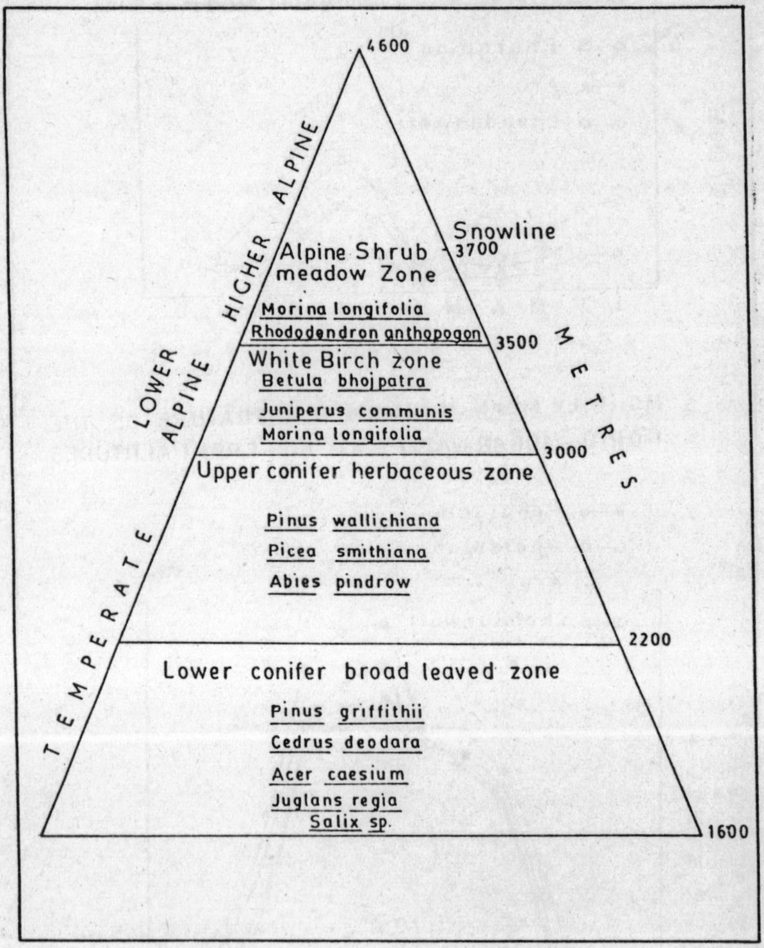

Fig. 7: *Altitudinal Succession of Vegetation in the Liddar Valley*

Lower Conifer Broad-leaved Zone

This zone extends up to the altitude of 2200 metres. In this zone *Pinus griffithii*, and *Cedrus deodara* grow along with broad-leaved species like *Acer caesium, Juglans regia* and *Salix* species. In addition many shrubby scrubs are associated.[8]

Upper Conifer Herbaceous Zone

The dominant arboreal vegetation of this zone includes *Pinus wallichiana, Picea smithiana, Abies pindrow* and several other evergreen species. This zone extends from Pahalgam (2200 metres) to Pissu Ghati and Liddarwat (3000 metres). The *Pinus wallichiana* and *Picea smithiana* constitute the major conifer group. They either occur in the form of pure colonies or are associated with broad-leaved trees. At Chandanwari some alpine flowers have been identified at an altitude of 2900 metres.[9]

White Birch Zone

The vegetation of this zone is observed from Zaijpal to Shishnag in the East Liddar and from Liddarwat to Satlanjan in the West Liddar valley. The dominant vegetation of this zone is *Betula bhojpatra*. It mainly occurs in the form of pure colonies. The prominent conifers are absent. The dominant grass of the meadows at Zaijpal and Liddarwat is *Morina longifolia*. The vegetation zone belongs to lower alpine belt and gradually merges with upper alpine elements when *Betula* species is replaced by *Juniperus* and *Rhododendron*.

Alpine Shrubs and Meadow Zone

This zone of vegetation is observed above 3500 metres altitude and is mainly localised near Shishnag and Kolahoi valleyhead. The extremely cold climate (Arctic type) helps in the growth of bushy vegetation of *Juniperus communis* and *Rhododendron anthopogon*. This vegetation gradually merges with the vast alpine meadows which abound in *Morina longifolia* and *Carduns nutans* grasses.

OBJECTIVE OF THE STUDY

The study region has polygenetic landforms produced by glacial and fluvial processes. They have produced specific landform features within different terrains. The objective of the present study is to analyse the glaciers and rivers and the features produced by them are as follows:

1. The identification of the landform features and the study of their

morphology, character, origin and chronology.

2. The majority of the features produced are interrelated and interdependent on each other. The objective is to study their interaction and from this discover the possible source of the material and thickness of the ice mass responsible for their formation. There will be an analysis of the interaction of the flow, the sediment and debris transport and bedform erosion. An attempt is made to investigate the relationship of process and form.
3. One of the major concerns is to investigate the mass balance of the existing glaciers, to study their movement and evaluate their earlier fluctuations records from the early photographs. The glacial fluctuations are closely related to the formation of the erosional and depositional landscape and are thus closely related to glacial geomorphology.
4. The last objective of the present study is to reconstruct the palaeogeographic glacial landscape from palaeobotanical, lichenometric and moraine morphological evidences.

METHOD OF STUDY

The study is based on detailed mapping of 1272 square kilometres of the area on the scale of 1 : 50,000. This is accompanied by systematic sampling. It has been achieved by extensive field observation and intensive laboratory work. The data collected from the field have been interpreted to derive the conclusion on geomorphological evolution of the region pertaining to its fluvial and glacial environments.

This work is mainly based on field observations and field measurements. The dimensions of the landform features were measured and photographs were taken in the field. The photographs provide panoramic views of various parts of the study area and evaluate the qualitative characteristics of the sediment fabrics. The quantitative analysis of sediment fabrics collected from different depositional features has also been conducted. The samples collected in the field have been analysed for sand, silt and clay fraction, by wet and dry sieving. The wet sieving has been used to remove particles in excess of a 2 millimetre diametre. The remainder of the particles of less than 2 millimetres size have been fractioned by dry sieving. The particles of large size of glacial deposits, particularly pebbles, and boulders have been analysed for long-axis orientation to evaluate their mode of deposition. The relative age of the deposits has been evaluated from boulder abundance, boulder frequency and weathering of the coarser sediments.

In order to understand the general characteristics of all the 17 existing glaciers, the two glaciers have been selected for the sample survey. They are the two largest glaciers of the area and each one is located in the East and the West Liddar valley. The mass balance records of the Shishram (the East Liddar) and the Kolahoi (the West Liddar) glaciers has been calculated from the accumulation and ablation measurements made during the winter and summer field seasons of 1983–85.

The relevant literature from hydrology, geology, palaeobotany, geomorphology and archaeology has been consulted to collect the secondary sources needed for the reconstruction of the palaeogeographic history of the area.

REVIEW OF EARLIER LITERATURE

Limited work has been done on the geomorphology of the Himalayan region. In particular the glacial landscape has received little consideration. There is good deal of information on the Himalayan glaciers *per se* in the Geological Survey of India publications but hardly on the glacial geomorphic features and processes. The general observations made by the early pioneers on the geography and the geology of the areas covered by glaciers have been found by the later, more scientific investigations, to be accurate.

The Kashmir Himalaya with its greater variety of landscapes has been investigated by a number of workers. The more significant contributions on the geology, geography and glaciology are discussed here. The application and relevance of the observation are discussed at appropriate places in the text.

Glaciology

The earlier reference on the glaciers of the Kashmir Himalaya was made by Schlagintweit brothers in the form of paintings and sketch maps.[10] Their paintings on Kolahoi glaciers were useful in measuring the glacier speed and glacierisation of the area. Holland classified the Kashmir glaciers into longitudinal and transverse types on the basis of their origin.[11] He gave only passing reference to the study area.

Neve was the first to study and prepare the map of the Kolahoi and its northern glaciers in detail.[12] He documented its movement during 1887 to 1909. His authentic observations were supported by later investigators.[13, 14]

Dainelli and De Terra were among the first few glaciologists to give a detailed account of glaciation of the Kashmir valley on the basis of the Pleistocene deposits. They detected four main phases of the glaciation on the basis of moraines and varves in Karewa deposits. Dainelli reported roche moutonnees and subglacial moraines along one of the valleywalls of southern part of the Liddar valley near Pahalgam but De Terra and Paterson did not accept his views on glacial origin.[15, 16]

Grinlinton mapped the glaciated section of the East Liddar valley and gave a general account of the distribution of few erosional and depositional features. On the basis of the depositional features he suggested that there were two glacial epochs. During the second epoch there were many advances and retreats.[17]

Wadia and Krishnan described the Pleistocene ice age deposits of Kashmir and also gave comprehensive stratigraphy of the southeastern part of Kashmir.[18, 19] They documented four glacial advances on the basis of glacial deposits.

Ahmad and Hashimi studied glacial history of Kolahoi (West Liddar). On the basis of the glacial deposits they documented three glacial advances. Ahmad also investigated the moraines of Kashmir and described their distribution.[20, 21]

Mayewski analysed the fluctuation records of the glaciers of the Himalaya and of the study region on the basis of the historical documents.[22] He also investigated the glacial chemistry of some of the Kashmir glaciers.[23]

Geology and Geography

The first systematic physiographic work in Kashmir particularly of the study region commences from the reconnaissance survey made by Lydekkar. He was the first to report on the geology and physiography of the area. He described a number of sections from Anantnag to Amarnath and from Pahalgam to Sind valley and the different rock formations and the structures of the area.[24]

Middlemiss gave a detailed stratigraphic and structural account of the area. He was the first to prepare a geological map of the area.[25] He divided the Silurian-Triassic sequence into two main groups. His findings were very accurate. Only a few revisions on them in respect to lithology have been made subsequently.[26, 27, 28] Gupta classified stratigraphic sequence of the area on the basis of the fossils. He classified Silurian-Devonian rocks of the study region.[29]

NOTES

1. Bhandari, N. and Nijampurkar, V.N. (1981). 'Glaciers: Our neglected water banks', *Science Today*, December, pp. 57-62, ref. p. 57.
2. Vohra, C.P. (1981). 'Himalayan glaciers', *The Himalaya: Aspects of Change*, Lall, J.S. and Moddie, A.D. (Eds.), Delhi, Oxford University Press, pp. 138-151, ref. p. 147.
3. Meher-Homji, V.M. (1971). 'The Climate of Srinagar and its Variability', *Geographical Review of India*, Vol. 33, No. 1, pp. 1-14.
4. Raza, M., Ahmad, A. and Mohamad, A. (1978). *The Valley of Kashmir: Volume I*, New Delhi, Vikas Publishing House, pp. 1-148, ref. p. 80.
5. Penck, A. (1973). 'Attempt at a Classification of Climate on Physiographic Basis, *Climatic Geomorphology*, Derbyshire, E. (Ed.) London, Macmillan, pp. 51-60.
6. Kachroo, P. (1985). 'A Note on Flora of Kashmir', *Geography of Jammu and Kashmir*, Husain, M. (Ed.), New Delhi; Ariana Publishing House, pp. 32-44, ref. pp. 37-39.
7. Sapru, B.L., Dhar, U. and Kachroo, P. (1975). 'Vegetation Studies in Jhelum Valley', *Botanique*, Vol. 6, pp. 151-164, ref. pp. 158-160.
8. See ref. 6, p. 38.
9. See ref. 6, p. 39.
10. Schlagintweit, H., Schlagintweit, A. and Schlagintweit, R. (1861). *Results of Scientific Mission to India and High Asia; with Atlas of Panoramas and views*, London: Leipzig, pp. 1-98, ref. p. 42.
11. Holland, T.H. (1907). 'A Preliminary Survey of Certain Glaciers in N.W. Himalayas — Introduction', *Record Geological Survey of India*, Vol. 35, Part 3, pp. 123-126.
12. Neve, E.F. (1910). 'Mt. Kolahoi and its Northern Glacier', *Alpine Journal*, Vol. 25, pp. 39-42.
13. Odell, N.E. (1963). 'The Kolahoi Northern Glacier, Kashmir', *Journal of Glaciology*; Vol. 4, pp. 633-635.
14. *Geographical Journal* (1925). 'The Movement of Indian Glaciers', Vol. 66, Part 4, pp. 331-333.
15. Dainelli, G. (1922). *Studi Sul glaciale: Spedizione Italiana de Filippi nell Himalaia Caracormume Turchestan Cinese (1913-14), Ser. II*, Bologona; Nicola Zanichelli, pp. 1-658.
16. De Terra, H. and Paterson, T.T. (1939). *Studies on the Ice Age in India and Associated Human Culture*, Washington D.C.; Publications Carnegie Institute, No. 493, pp. 1-493.
17. Grinlinton, J.L. (1928). 'Former glaciation of the East Liddar Valley, Kashmir', *Memoir, Geological Survey of India*, Vol. 49, Part 2, pp. 289-388.
18. Wadia, D.N. (1941). 'Pleistocene Ice Age deposits of Kashmir', *Proceedings of the National Institute of Sciences of India*, Vol. 7, pp. 49-60.

19. Krishnan, M.S. (1960). *Geology of India and Burma*, Madras: Higginbothams Pvt. Ltd., pp. 1-603, ref. pp. 563-67.
20. Ahmad, N. and Hashimi, N.H. (1974). 'Glacial History of Kolahoi Glacier, Kashmir, India', *Journal of Glaciology*, Vol. 13, pp. 279-283.
21. Ahmad, N. (1979). 'Morainic Deposits in Kashmir Himalayas', *Moraines and Varves: Origin/Genesis/Classification*, Proceedings of an INQA Symposium on Genesis and Lithology of Quaternary deposits, Schluchtter, Ch. (Ed.), Rotterdam; A.A. Balkema, pp. 59-63.
22. Mayewski, P.A. and Jaschke, P.A. (1979). 'Himalayan and Trans Himalayan Glacier Fluctuations Since A.D. 1812', *Arctic and Alpine Research*, Vol. II, No. 5, pp. 267-287.
23. Mayewski, P.A. and Ahmad, N. (1981). 'Reconnaissance Glacio-chemical Studies in the Indian Himalayas', Proceedings of *Eastern Snow Conference*, 38 annual meeting, p. 48.
24. Lydekkar, R. (1883). 'Notes on Geology of Kashmir, Kishtwar and Pangi', *Memoir, Geological Survey of India*, Vol. 22, pp. 1-108.
25. Middlemiss, C.S. (1911). 'Section in Pir Panjal Range, and Sind Valley', *Record, Geological Survey of India*, Vol. 41 Part 2, pp. 1-30.
26. Shah, S.K. (1972). 'Stratigraphic Studies on Lower Palaeozoic Sequence of Anantnag District, Kashmir', *Himalayan Geology*, Vol. II, p. 477.
27. Bhat, D.K. (1975). 'On the Quaternary Geology of Kashmir valley with special reference to stratigraphy and sedimentation', *Geological Survey of India*, Miscellaneous Publication, No. 24, pp. 188-203.
28. Pascoe, E.H. (1959). *A Manual of the Geology of India and Burma*, Vol. II, Calcutta; Manager, Government of India Press, pp. 485-1343.
29. Gupta, V.J. (1966). 'Palaeontology, Stratigraphy and Structure of the area Southeast of Srinagar', Unpublished Ph.D. thesis, Panjab University, pp. 1-285.

The Shale-Slate-Greywacke formation is overlain by pale, pink or greyish coloured quartzite near the localities of the Hapatnar, Shumal and Riatang (Fig. 8).

Orthoquartzite-Carbonate Group

The Shale-Slate-Greywacke Group is overlain by the thick pile of quartzites and sandstone of various types intercalated with limestone beds. The exposures are observed in the localities of Nagbal, Driyan and the easternmost parts of the Kanjit, Gurdeman, Aishmaqam and Kolahoi-Basmi anticline (Fig. 8). This group has two horizons, one comprising of Muth Quartzite and the other of Syringothyris Limestone. The rocks are mostly hard, compact and massive. The Muth Quartzite contains thick sequence of quartzite and quartzitic sandstone. They are 5 to 6 metres thick near Aishmaqam. The quartzite is closely jointed, while schists are highly crushed and pulverised. They have been dated to Devonian age.[14]

Muth Quartzite formations are overlain by 150 metres thick Syringothyris Limestone near Kotsu and Aishmaqam. It is not composed of limestone alone as the name indicates, but it is limestone interbedded with quartzite and shale. Gupta divided this zone into three distinct groups on the basis of fossils and dated it to Lower Carboniferous.[15, 16, 17] But the main difference in lithology is the degree of dolomitisation. Though the limestones of this zone are massive they are highly jointed (Figs. 9B and 10C). They strike northwest-southeast. The Muth Quartzite is overlain by thick, grey, hard, quartzites, sandstones and shales at Pindobal-Ziral section. They are moderately jointed. This formation represents Passage bed of Middlemiss. Some geologists observed that there is no Passage bed between Syringothyris Limestone and Fenestella Shale as seen in the locality of Ayemi and in Pindobal-Ziral section, where it occurs between shale and slate. Similar situation exists in the localities of Kollur, Salar and Kolser Passage beds between Muth Quartzite and Syringothyris Limestone.[18]

Outer Volcanic Group

The Orthoquartzite-Carbonate Group is conformably overlain by a set of rocks having complex petrological characters. They are pyroclastic in nature and are named as Agglomeratic Slates. The rocks show a northwest-southeast strike trend. They dip 30° to 40° towards northeast.

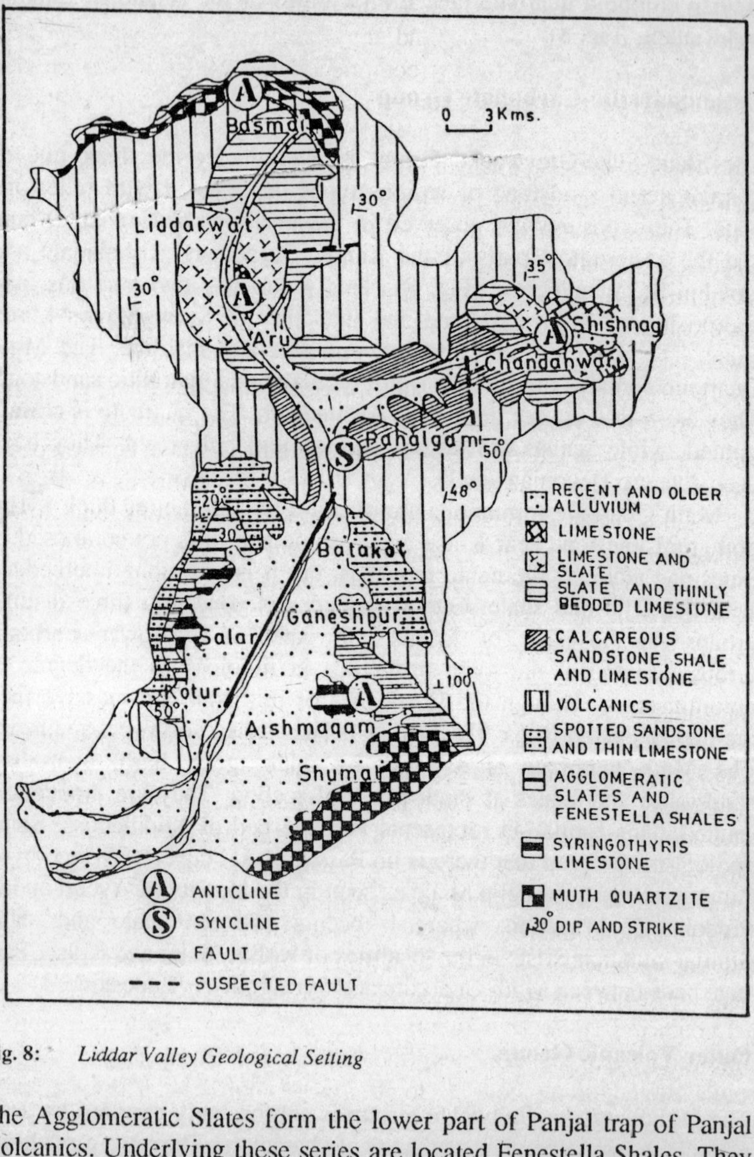

Fig. 8: *Liddar Valley Geological Setting*

The Agglomeratic Slates form the lower part of Panjal trap of Panjal Volcanics. Underlying these series are located Fenestella Shales. They have considerable thickness of 610 metres and are composed to black, sandy shale. Wadia too has clubbed this rock series with overlying

Volcanic Agglomeratic Slates.[19] The Fenestella Shales and Agglomeratic Slates have been included in the volcanic group because the rocks are mostly crystalline and they weather and erode relatively slowly. The Fenestella Shales comprise shales ranging from greyish white to greyish black. The quartzites are found along these shales. These deposits have been regarded as Carboniferous in age. The deposits are exposed along the Yanar. The Fenestella Shales are overlain by shales, slates, sandstones and quartzites. These deposits are apparently heterogeneous but are similar in their pyroclastic origin. The formation is commonly called Agglomeratic Slates.[20] This formation is well preserved in the localities of Batakot, Parangam, Kullur, Naubug and Kolahoi-Basmai anticline. This formation gradually thins out towards the northeastern part of Kolahoi.

The Agglomeratic Slates show regular gradation from vitric to lithic type. The deposits become more conglomeratic at the top. The formation is almost barren but recently some fossils have been reported near the localities of Nishnalnar and Virsiran. On the basis of the fossils it has been dated between middle Carboniferous and lower Carboniferous.[21, 22]

Classic Volcanic Group (Punjal Volcanics)

Agglomeratic Slates are overlain by a thick homogeneous mass of lava, known as Panjal Volcanics. The rocks are green and cover a major portion of the northern part of the Liddar valley. They are compact, thickly bedded and massive in nature. They are very hard and moderately to highly jointed. The Panjal Volcanics cover the high mountain ridges and valleywalls of the East and the West Liddar valley. The joints have spacing of 1/2 to 1 metre. The Panjal Volcanics forms the towering cliffs of the area.

The name of the rock has been derived from the Pir Panjal range in the Kashmir which is primarily covered by volcanic traps. The volcanic rocks are medium to fine-grained and compact in nature. The upper contact of the rocks with the Zewan Series is seen near Pissu Ghati. The beds have a strike, near the contact, of 55° northwest to 55° southeast and dip of about 30° to 60° towards northeast. The lower contact of the volcanics is observed near Pahalgam. The geologists of the Geological Survey of India have observed prominent joint patterns in the volcanics. The majority of the glacial erosional landform features have been carved in Panjal Volcanics because the rocks are very hard and are resistant to erosion. The volcanics have taken the markings of glacial abrasion very well and they are well preserved as striae. The cirques have been carved in volcanic surfaces along the joints at higher altitude. Similarly different stairways have been produced in glacial troughs near the contact zone of volcanics with different lithologies. The volcanics have usually developed steep slope and high cliffs as they are very resistant to erosion due to crystalline nature of the rock

Fig. 9A: *Massive Thickly Bedded Panjal Volcanics (A) near Liddarwat Forming the Vertical Cliff*

Fig 9B: *Highly Jointed Syringothryis Limestones (A) near Aishmaqam. Thin inter-Bedded Shale Parting (B) Separates Limestone beds*

and amygdaloidal structure (Figs. 9A and 10E). Near Ganshibal and Pahalgam the volcanics have developed narrow gorges on the Liddar river. The volcanics have been regarded as Permo-Carboniferous in age.[23, 24, 25]

Sandstone Shale Group

The volcanic group is conformably overlain by complex set of rocks being composed of sandstones, siliceous limestones and shales. The sandstones and limestones are massive and tough while shales are soft and highly cleaved. The chief formations of this group are Burzulpathar Series and Zewan Series. Burzulpathar Series contain siliceous and carboniferous shales, black and grey sandstones and few limestones.[26] The formation lies directly over Panjal Volcanics. The exposures are observed along channel sites. In the localities of Burzulpatharnar, and Minmargnar the formation consists of grey shales. The shales are very soft and easily erodable. Broad valleys have been produced in them by glacial and glacio-fluvial erosion. The major portion of (morainic) soil has been formed from this rock group. This group is overlain by fossiliferous shales, limestones and calcareous sandstones. The beds are of Lower Triassic age and are called Zewan Series.[27, 28] They have been named after the village Zewan where they were first observed. The formation covers the maximum area among the sedimentaries. The sandstone beds of Zewan Series strike at 20°

Fig. 9C: *An Anticlinal Fold (A) near Zaijpal, the Central Part of the Fold Comprises Calcareous Sandstones*

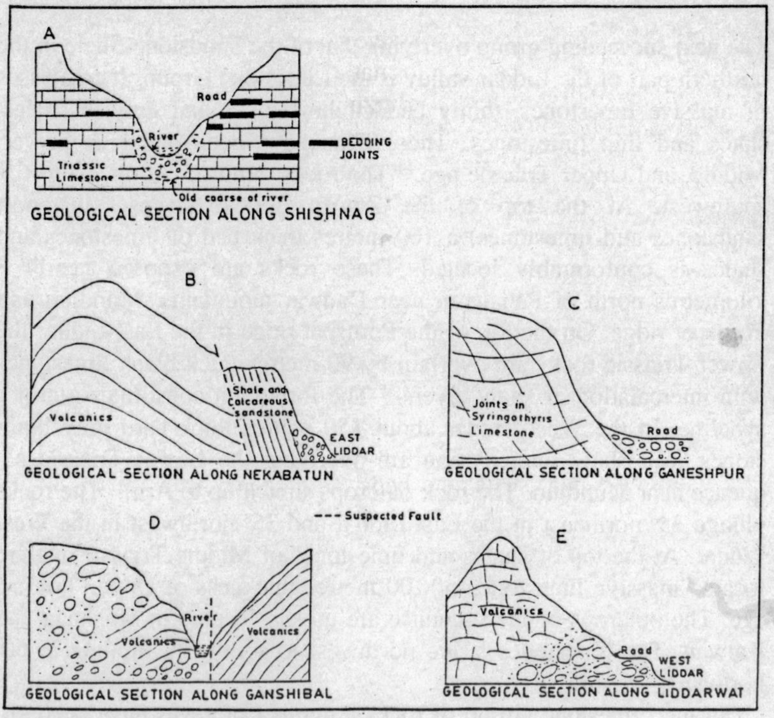

Fig. 10: *Geological Sketches of the Selected Sites of Liddar Valley*

northwest to 20° southeast and dip 60° to 65° northwest. The rocks are highly jointed and the joints have spacing of 1/2 to 2 metres. At the base of the formation cherry limestone and grey quartzites are exposed around the localities of Pissu Ghati and Niliken Gur. At the upper side their contact is with limestone. The formation generally occurs along the flanks of the valley. They stretch from south to north along the West and the East Liddar river from Mamal to Liddarwat and Pahalgam to Chandanwari respectively. The thick outcrops of shales and sandstones are also exposed in the localities of Athnar and Burzulpathar areas. The sandstones and shales have been the main source region of the material of most of the glacial depositional features. The broad glacial troughs are generally located in this rock group and reflect the severe erosion of the soft rocks.

Limestone Group

The next succeeding group overlying that of the Sandstone-Shale in the northern part of the Liddar valley is the Limestone Group. It comprises of massive limestones, thinly bedded limestones and greyish shales, slates and thin limestones. These beds have been dated to Lower, Middle, and Upper Triassic age.[29] The rocks trend 65° northeast to 65° southwest. At the top of the Zewan Series (shales, siltstones, sandstones and limestones) a 100 metres thick bed of limestones and shales is conformably located. These rocks are exposed nearly 3 kilometres north of Pahalgam near Dadwar mountains, Mundlun and Poruspat ridge. On the top of the Poruspat ridge in the East Liddar the Lower Triassic rocks are overlain by 90 metres thick black limestones with intercalation of shaly layers.[30] The formation constitutes a minor syncline. In the West Liddar about 350 metres thick hard grey limestones with shaly intercalation are overlying the Lower Triassic sequence near Mundlun. The rock outcrops stretch up to Aru.[31] The rocks plunge 25° northeast in the East Liddar and 35° northwest in the West Liddar. At the top of shales and limestones of Middle Triassic age are located massive limestones of 100 metres thickness of Upper Triassic age. The outcrops of the sequence are mostly located in Shishnag and Kalwan. The limestones strike northwest to southeast, dipping at 60° northeast.

In the Limestone Group of rocks a number of caves have been observed above the present water level near Shishnag basin. Initially, the caves were formed by the mechanical erosion of the river which exploited the joints of the limestone bed. Later they were enlarged by solution action (Fig. 10A).

A number of waterfalls and cascades have been observed in the limestone beds exposed along the course of the river. The surface of the limestone has produced riegels between Zaijpal-Shishnag and Liddarwat-Satlanjan sections of the East and the West Liddar respectively. Also a number of large cirques and cirque lakes have been produced in the limestones by the Liddar glaciers.

Glacial and Fluvial Alluvium Group

These deposits occur in the form of older and newer alluvium. The older alluvium is deposited in the form of moraines, eskers, kames and the scree. The older alluvium overlies the older rock groups from the

Shishnag to Ganshibal and Kolahoi to Pahalgam along the valley sections of the East and the West Liddar respectively. They are Subrecent (Late Pleistocene) in age. The recent alluvium brought from the East and West Liddar rivers has been deposited along the river beds and banks in the form of silt, boulders, and pebbles.

MAJOR STRUCTURES OF THE AREA

The major rock groups of the Liddar valley are folded into anticlines and synclines. The syncline axis passes near Pahalgam whereas the corresponding anticlines are in north and south of Pahalgam (Fig. 8). The Liddar valley anticline is the famous one and was first identified and mapped by Middlemiss.[32] It has normal bedding dips on both the limbs in opposite directions. The central part of this structure has been considerably denuded. As a result of this a large part of rocks have been exposed. The following major structures have been identified by the geologists.[33, 34, 35, 36, 37]

Anticline

The famous Liddar valley anticline is located in the central part of the study area. It has a northwest-southeast strike and steep plunge toward the north-northwest.[38] The central part of the anticline is located near Manzgam spur comprising of thinly bedded siltstones and shales. It is overlain by micaceous sandstones, greywackes and shales (Fig. 11). This formation forms the southern limb of the anticline of the south of Aishmaqam. The rocks near Kotsu show dip of 45° due southwest. The northern limb of the anticline is primarily composed of quartzite. These rock beds have uniform dip of 30° northeast towards Pahalgam. Near Pahalgam the anticline joins with the core of the syncline. The core of the syncline is primarily composed of calcareous sequence of Sandstone-Shale Group.[39] The corresponding anticlinal axis in the northern part of Liddar valley runs along Chandanwari and Aru. The limbs of the anticline are composed of rocks of Zewan Series which strike in a north-northwest to south-southeast direction with the dip varying from 40° to 70° towards northeast and northwest. The anticlinal folds form the northern limbs of the Pahalgam syncline. In the northwestern part of Liddar valley a Basmai-Kolahoi anticline runs in northwest direction. The limbs of the anticline are composed of rocks of Muth Quartzite and Agglomeratic Slates which strike in northwest.

The oldest formation (Muth Quartzite) are exposed in the core of the anticline.

Faults

A number of faults have been identified but the major one runs from north-northeast to south-southwest. It has been recognised near Kotsu and Batakot. The fault runs south of Pahalgam near the Liddar river. As a result of faulting the younger Agglomeratic Slates come in juxtaposition with the older shales. The displacement of the fault is nearly 1000 metres and considerable crushing of rock has been noted near Geneshpur and Kotsu. The geologists of the Geological Survey of India have suspected a fault near Nekbatun at the contact of Panjal Volcanics with Zewan Formation. There has been displacement of beds for nearly 1500 metres. The crushing of rocks have been observed

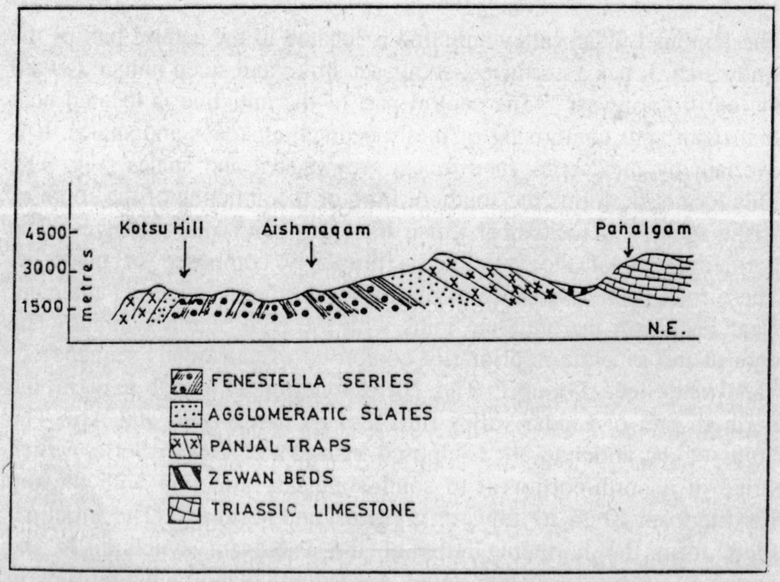

Fig. 11: *Section Along the Liddar Valley Anticlines, Kashmir*

such of nekbatun (Fig. 10B). Another minor fault has been observed east of Liddarwat.

Joints

Prominent joints have been observed by the author and the geologists of the Geological Survey of India in the northern part of Liddar valley along the Panjal Volcanics and Sandstone-Shale Group of rocks. They are appended in Table 3. The spacing of Joints ranges between 25 centimetres and 2 metres in Sandstone-Shale Group (Table 3). In volcanics the spacing of joints range between 50 centimetres and 1 metre (Fig. 10).

Table 3: *Joint Patterns in the Rock Types of Selected Sites of Liddar Valley*

Strike	*Amount of dip*	*Dip direction*	*Spacing of Joints*
Liddarwat-Zaijpal (Zewan Series)			
N 40°W - S 40°E	65°	NE	25 cm - 1 metre
N 30°W - S 30°E	15°	SE	1 metre - 2 metres
N 80°W - S 80°E	65° - 80°	SW	50 cm - 1 metre
N 60°E - S 60°W	45°	SE	25 to 40 cm
N 10° - 25°W - S 10°- 25°	35°	SE	filled in the quartz veins 1 - 3 metres apart
Aru-Chandanwari (Volcanics)			
N 45°E - S 45°W	65° - 70°	NW	80 cm - 1 metre
N 35°W - S 35°E	85°	NE	75 cm
Near Pahalgam between Laripora (East Liddar) and Mandlun (West Liddar)			
N 40° - 50°W to S 40° - 50°E	20° - 30°	SW	1/2 metre - 1 metre
N 60°W - S 60°E	45°	SW	tight to slightly open
N 70° - 75°E to S 70° - 75°W	55°	NW	1 metre

NOTES

1. Middlemiss, C.S. (1910). 'A Revision on the Silurian-Triassic sequence in Kashmir', *Record, Geological Survey of India*, Vol. 40, Part 3, pp. 206-260.
2. Kapoor, H.M. (1979). 'Gondwana of Kashmir: A Reappraisal', *Fourth International Gondwana Symposium, Calcutta, India*, 1977, Delhi : Hindustan Publishing Corporation, Volume 2, pp. 443-462.
3. Wadia, D.N. (1937). 'Permo-Carboniferous in the Tertiary Zone of Kashmir Himalayas', *Record, Geological Survey of India*, Vol. 48, Part 2, pp. 1-16, ref. p. 5.
4. Shah, S.K. (1972). 'Stratigraphic studies of Lower Palaeozoic sequence of Anantnag district, Kashmir', *Himalayan Geology*, Vol. 2, pp. 468-80.
5. Fuchs, Gerhard, (1975). 'Contribution to the geology of NW Himalayas', *Eigentumer Heraugeber and verleger : Geologisches Bundenstalt*, Wien 32, pp. 1-59.
6. Gupta, V.J. (1969). 'Palaeozoic Stratigraphy of the area southeast of Srinagar', *Research Bulletin, Panjab University*, Vol. 20, part (I-II), pp. 1-14.
7. Middlemiss, C.S. (1911). 'Section in Pirpanjal Ranges and Sind Valley, Kashmir', *Record Geological Survey of India*, Vol. 41, Part 2, pp. 1-30.
8. Gupta, V.J. (1973). *Indian Palaeozoic Stratigraphy*, Delhi : Hindustan Publishing Corporation (India), ref. pp. 34-67.
9. Wadia, D.N. (1934). 'The Cambrian-Triassic Sequence of North-Western Kashmir (parts of Muzzafarabad and Baramula Districts)', *Record, Geological Survey of India*, Vol. 68, Part 2, pp. 121-176.
10. Srikantia, S.V. and Bhargava, O.N. (1983). 'Geology of the Palaeozoic sequence of the Kashmir Tethys Himalayan basin in the Liddar Valley, Jammu and Kashmir', *Geological Society of India*, Vol. 24, No. 7, pp. 363-377.
11. Sastri, C.G.K. (1961). 'Report on Systematic geological Mapping in Liddar Valley, Anantnag district, Kashmir', *Geological Survey of India*, Misc. Pub. No. 2., pp. 33-39.
12. Gupta, V.J. (1971). 'Silurian System of India subcontinent', *Research Bulletin of Panjab University*, Vol. 22, Part III - IV pp. 339-343.
13. See ref. 10, ref. pp. 366-367.
14. Gupta, V.J. (1970). 'Muth Quartzite Fauna from Kashmir, India', *Research Bulletin, Panjab University*, Vol. 22, Part 1-2, pp. 1-22.
15. See ref. 12, pp. 341-42.
16. Middlemiss, C.S. (1909). 'Gondwana and related marine sedimentary system of India', *Record, Geological Survey of India*, Vol. 37, Part 4, ref. pp. 321-22.
17. Pascoe, E.H. (1959). *A Manual of the Geology of India and Burma*, Vol. II, Calcutta : Manager, Government of India Press, pp. 485-1343, ref. pp. 637-641 and 738-742.
18. Sharma, V.P. and Sehgal, M.N. (1976). 'Some of the new Observations on Stratigraphy and Structure of Liddar Valley anticline, Kashmir', *Geological Survey of India, Kashmir Circle*, No. 14, pp. 56-66.
19. Wadia, D.N. (1953). Geology of India, London; Macmillan & Co. Ltd., pp. 1-531, ref. pp. 216-227.
20. Pareek, H.S. (1977). 'On the studies of Agglomeratic Slates and Panjal traps in Jhelum, Liddar and Sind Valley, Kashmir', *Record, Geological Survey of India*,

Vol. 107, pp. 12-37.

21. Bion, H.S. and Middlemiss, C.S. (1926). 'The Fauna of Agglomeratic Slate Series of Kashmir', *Geological Survey of India*, Vol. 12, pp. 1-55, ref. pp. 24-27.
22. Gupta, V.J. (1966). 'Palaeontology, Stratigraphy and Structure of the area southeast of Srinagar', *Unpublished Ph.D. Thesis*, Panjab University.
23. Krishnan, M.S. (1960). *Geology of India and Burma*, Madras: Higginbothams, Pvt. Ltd., pp. 1-604, ref. pp. 354-358.
24. See ref. 17, pp. 738-742.
25. See ref. 3, p. 5.
26. Ghosh, D.K. (1973). Report on Liddar Valley Hydel Section', *Geological Survey of India, Special paper*, No. 5, pp. 17-21.
27. See ref. 10, p. 373.
28. Nakazawa, K., Kapoor, H.M., Ishi, K., Bando, P., Okimura, Y.K. and Tokuoka, T. (1975). 'The Upper Per mian and Lower Triassics in Kashmir, India', *Memoir of Faculty of Science, Koyoto University, Serial Geology-Minerology*, Vol. 42, pp. 1-106.
29. See ref. 26, p. 21.
30. Singh, D.P. (1974). 'A Report on the Systematic Geological Mapping of Liddar Valley, Anantnag district, Jammu and Kashmir', *Geological Survey of India*, No. 34, pp. 47-58.
31. Hashimi, N.H. (1971). 'Notes on the Geology of West Liddar Valley, Kashmir', *Himalayan Geology*, Vol. 1, pp. 284-287.
32. See ref. 1, pp. 257-260.
33. Koul, B.L. (1976). 'Structures of Liddar Valley Anticline'. *International Seminar on Himalayan Geology*, New Delhi, p. 59.
34. Gansser, A. (1964). *Geology of the Himalayas*, New York; John Willey and Sons. pp. 1-289, ref. pp. 49-51.
35. See ref. 4, pp. 477-478.
36. Fuchs, G. and Gupta, V.J. (1971). 'Palaeozoic Stratigraphy of Kashmir, Kishtwar and Chamba Regions', *Verhandlungen der Geologisches Bundenstalt in Wien*, No. 1, pp. 68-97.
37. See ref. 5, pp. 18-20.
38. See ref. 10, pp. 374-375.
39. See ref. 1, pp. 2-57.

Dynamics of the Liddar Glaciers

PATTERN AND EXTENT OF THE EXISTING GLACIERS

The existing glaciers of the Liddar valley are the relics of their mighty predecessors formed during the Pleistocene epoch. Presently they are located at higher altitudes ranging between 3700 and 5000 metres, to which they have retreated during the Holocene epoch. They have descended to an altitude as low as 2000 metres (above mean sea level) during the last maximum glaciation. There are 17 glaciers in the study region. They range in length from 0.1 to 6 kilometres and cover an area of 50 square kilometres. The Kolahoi and Shishram are two major glaciers located at an altitude of 5000 and 4950 metres respectively. They are 5 to 6 kilometres long and extend along Kolahoi and Kunyirhayan mountains (Fig. 12). The smallest glaciers are less than 10 metres long and are located in the sheltered cirques of mountain flanks. Some of the neve fields of these glaciers coalesce with each other and join the main icefence to the north and to the east. A few of the glaciers such as Gratbal are so small that they lie in the niches which they have sculptured. Ten glaciers are longer than 1 km. All are located in higher parts of mountain ranges in the North and Northeast of Liddar river lineament. They terminate at 3700 to 4000 metres above mean sea level.

The present distribution of glaciers is notably asymmetric. A majority of them flow toward north and northwest. The north flowing glaciers are between 2 to 6 km long and their south flowing counterparts are less than 2 km long (Table 4). Most of the glaciers are located in cirque basins situated along the precipitous slopes of mountain ranges of Kolahoi, Kunyirhayan and Katsal. The aretes and cols forming the ridges behind the cirques reveal that ice was never thick enough in the Pleistocene epoch to cover the highest parts of the mountain ranges.[1] Most of the higher peaks of the mountain are too steep to retain winter snow on their upper slopes. Ice from the peaks descended into the cirque basins and trenches at lower altitudes.[2] The higher

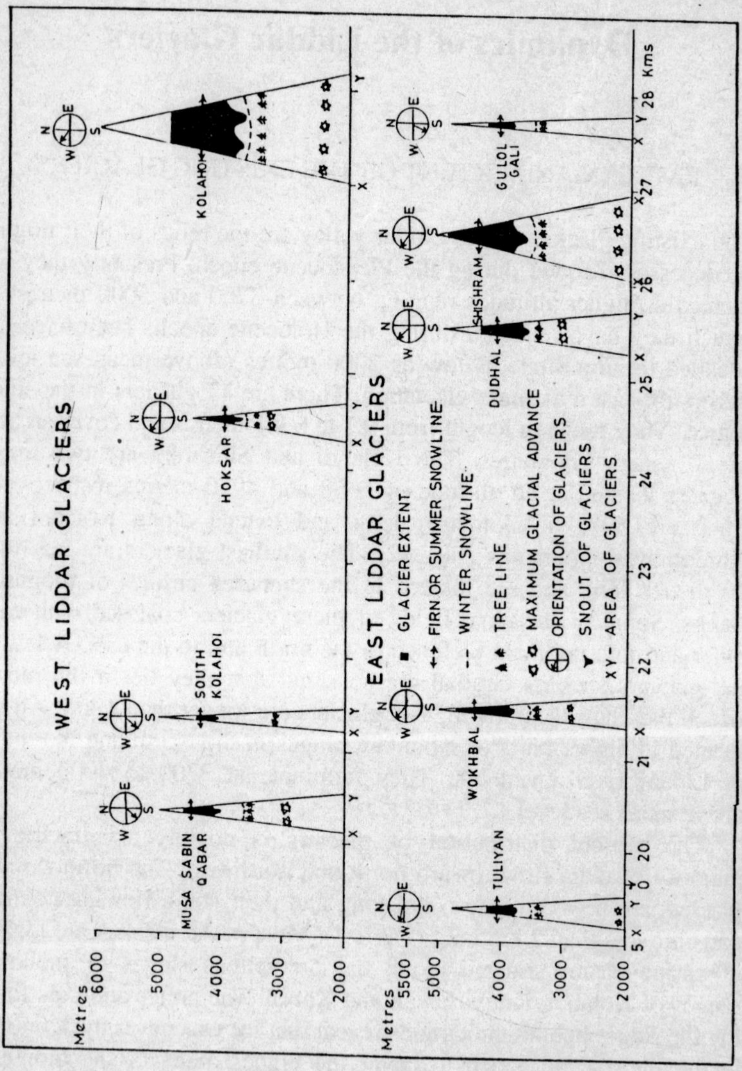

Fig. 12: *Horizontal and Vertical Zonation of Liddar Glaciers*

Table 4: *Altitudinal Extend and Relative Relief of the Liddar Glaciers*

Glacier	*Area (km²)*	*Direction of snout from the catchment area*	*Length (km)*	*Breadth near snout (metres)*	*Extension of glacier height*		*Relative relief (metres)*
					Snout (metres)	*Source Area (metres)*	
1	2	3	4	5	6	7	8
West Liddar/ Kolahoi Group							
1. Kolahoi	11.9	Northwest	6.5	321	3690	5000	1310
2. Musa Sabin Qabar	3.0	Southwest	1.8	130	3950	4800	850
3. Hoksar	2.4	Northwest	1.4	110	3600	4200	600
4. South Kolahoi	1.4	Southwest	0.6	85	4000	4200	200
5. Handut (1)	1.5	Northeast	0.7	90	3900	4400	500
6. Handut (2)	2.4	Northeast	1.1	100	3950	4400	450
East Liddar/ Shishram Group							
7. Shishram	9.9	Northwest	4.5	226	3740	4900	1160
8. Golol Gali	0.7	Northwest	0.3	60	4100	4600	500
9. Gratbal	0.8	Northwest	0.2	75	4000	4600	600
10. (A)	0.5	West	0.1	40	4000	4600	600
11. (B)	0.4	West	0.2	50	4100	4600	500
12. (C)	0.3	West	0.1	40	4100	4600	500
13. (D)	0.5	West	0.2	70	4100	4600	500
14. Son Sar	6.0	Northwest	2.5	230	3900	4700	800
15. Dhudhal	3.8	Northwest	1.6	200	3800	4800	1000
16. Tuliyan	4.1	Northwest	1.2	200	3700	4600	900
17. Manipal	2.6	Northwest	1.1	95	4100	4800	700

Source: Field Observations and Topographical Maps.

slopes being devoid of ice mass were thus exposed to frost action and lower slopes and valley basins to glacial action, resulting in the formation of steep inclined upper slope and polished parabolic lower slopes. The glacial action has produced bench cuts at the junction of the upper steep and the lower parabolic slopes on the hard volcanic rocks at altitudes of 3500 and 4200 metres in the East and the West Liddar valley.[3,4]

NATURAL DETERMINANTS OF GLACIERS

The elevation of the ridges, orientation of slope, the amount of winter precipitation and characteristics of summer temperature, determine the site, regional pattern and limit of glacierisation.

Elevation Aspect

The extent and condition of glaciers depend primarily on critical elevations which bring larger areas above snowline. The Liddar glaciers have their origin and termination at critical elevations of 3700 and 5000 metres above the mean sea level. The general rise in elevation at which the glaciers exist towards the north and the northeast is a response to northward decrease in temperature modulated by heavy snowfall. The upliftment of Liddar basin was effected during the Late Miocene/Early Pleistocene Epoch.[5,6] The fluvial processes could not keep pace with the rapid uplift when the elevations increased above the glacierisation limit. The high elevations promoted low temperature and abundant snowfall. The glaciers occupied the protected topographic situations of the northern part of the Liddar valley. Their winter accumulation was somewhat greater than the surrounding areas subject to wind drifting. In summer they received reduced insolation because of rain-shadow effect.[7,8,9] In contrast, some of the neighbouring mountain summits and slopes of west and central Liddar valley were devoid of ice (Fig. 13).

The annual snowline varies from one glacier to the other in the Liddar valley and ranges between 3700 and 4000 metres (Fig. 12). In the vicinity of Kolahoi Ganj and Shishnag lake at an altitude of 3400 metres, patches of perennial ice are observed. They are located in deep gorges at a depth of 200 metres below the maximum glaciation limit. The ice patches reflect the possible advance of glacier during winter season followed by minor retreat. During the retreat the ice patches got detached from the main glacial tongue due to orographic barriers.[10]

Therefore the actual winter snowline of Kolahoi and Shishram glaciers during 1983-84 stood at an altitude of 3400 metres. In the Pleistocene epoch it advanced to an altitude of 2000 metres.

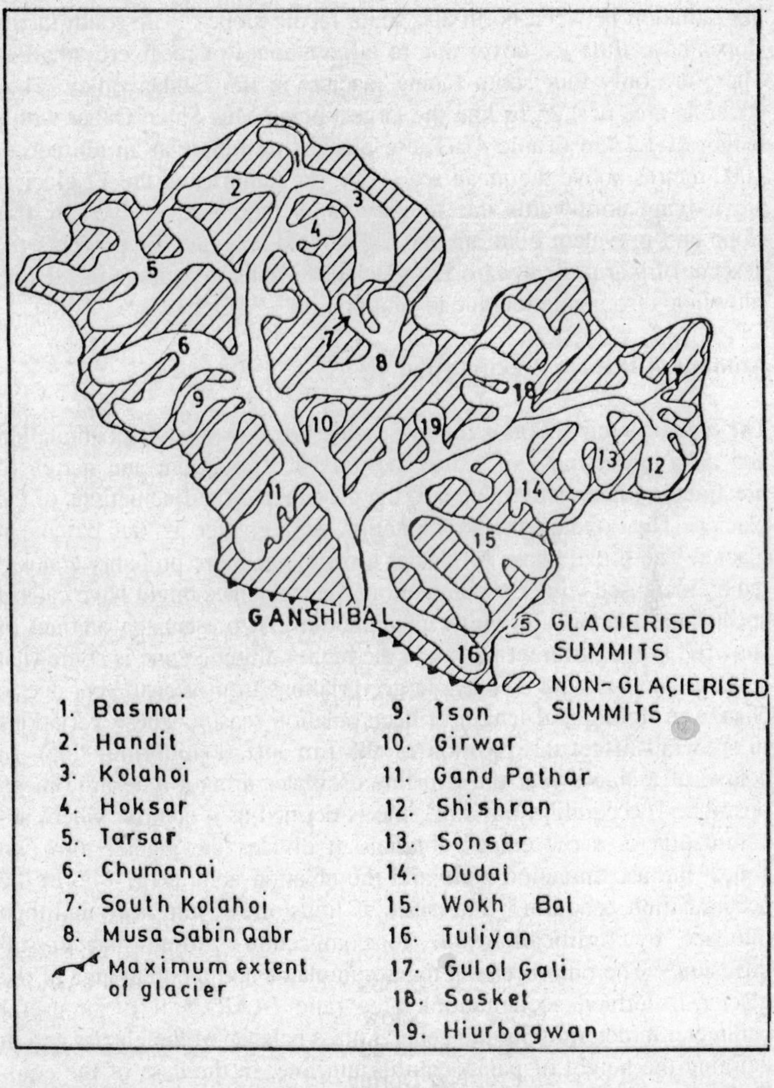

Fig. 13: *Glacier Extent : Pahalgam-Ganshibal Stade*

Orientation of Slope

The orientation of slope plays a dominant role in the growth of a glacier. This is partly due to wind drifting of snow into high accumulation basin, but it is more effectively related to the difference in incoming radiation between north and south facing slope.[11] The south facing slopes have little ice cover due to large amount of received radiation. There are only four south facing glaciers in the Liddar valley. They cover an area of 4.25 sq km, the largest being Mus Sabin Qabar with a length cf 1.7 km (Table 4). These glaciers terminate at an altitude of 4000 metres above the mean sea level. The majority of the 17 glaciers are moving northwards due to very small radiation received by this slope and prevalent cold air mass. The cold stagnant air prevents the descent of warm air toward the glacier and thus insulates the existing snowfields from ablation due to sublimation[12, 13] (Fig. 12).

Amount of Winter Precipitation

The winter precipitation is the most widespread means of accumulation and adds larger mass of ice to the glacier. The extent and nature of precipitation is directly related to the growth and the fluctuations of the glacier. The greater the precipitation, the greater is the extent of glacier. The initial phase of glacier growth was more probably controlled by increased winter accumulation. Such changes might have caused spectacular increase in winter precipitation. At present the amount of snowfall at the different places in the higher altitude zone is somewhat similar. The snowfall displays large variations from year to year due to change in timing and length of accumulation season. These variations in snowfall affect the position of the firn line (equilibrium line). In course of a single year the firn line oscillates around a definite mean position. The equilibrium (firn) line is defined as a contour where accumulation of snow equals ablation. It divides the glacier into two zones: the accumulation zone and the ablation zone (Fig. 14). In the accumulation zone the winter snow is transformed into neve and then into ice by firnification due to compaction settling and crystal diffusion.[13] The ratio between the accumulated and the total area of the glacier is termed accumulation area ratio (AAR).[14] It is an useful parameter in determining the general mass balance of the glacier and in defining the height of palaeoequilibrium line. In the case of the equilibrium glacier the accumulation area ratio varies between 0.6 and 0.7.

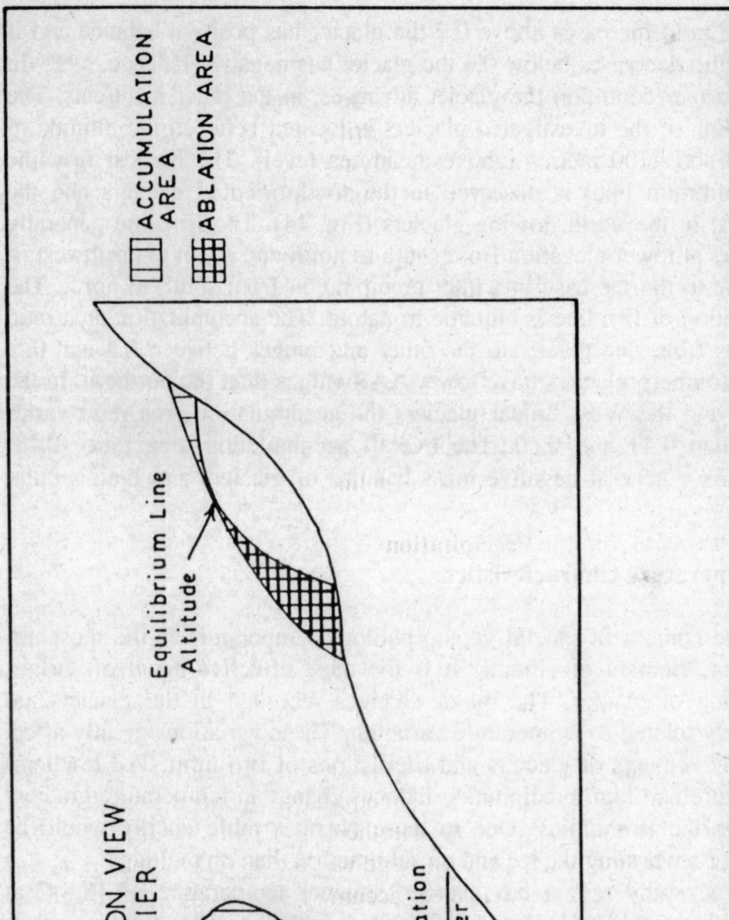

Fig. 14: *Longitudinal Profile and Cross-section of Liddar Valley Glacier*

If the ratio increases above 0.7 the glacier has positive balance and if the ratio decreases below 0.6 the glacier has negative balance.[15 and 16] In the former condition the glacier advances, in the latter it retreats. The firn line of the investigated glaciers is located between the altitude of 4400 and 4700 metres (above mean sea level). The highest firn line (equilibrium line) is observed in the south-oriented glaciers and the lowest in the north-flowing glaciers (Fig. 14). The firn line generally occurs at lower elevation from south to north and south to northwest. It is due to the increase in winter precipitation from south to north. The elevation of firn line is climatic in nature. The accumulation area ratio varies from one glacier to the other and ranges between 0.4 and 0.5. The southern glaciers have lower AAR values than the northern. In the East and the West Liddar glaciers the accumulation area ratio varies between 0.47 and 0.50. The overall accumulation area ratio (0.48) reveals a general negative mass balance of glaciers and their secular retreat.

Temperature Characteristics

In the context of glacial geomorphology temperature is the most important element of climate. It is the most effective agent of surface melting of glacier. The major changes recorded in the glaciers are closely related to temperature variation. These variations greatly affect the movements of glaciers and oscillations of firn limit. The low temperature and high precipitation induces change in temperature gradient above the firn limit.[17] Due to these changes more energy would be spent on warming the ice and on sublimation than on melting.[18]

The study region has average summer temperature of 18.8°C at Chandanwari. The high temperature raises the basal ice temperature in the glaciers to melting point from snout to firn limit and surface ice temperature to 0°C. Due to higher summer temperature the glaciers melt and lose their mass. They retreat due to negative mass balance. Presently the snouts of the glaciers exist at the mean elevation of 3700 metres above the mean sea level, but during the Pleistocene epoch the ice bodies advanced to 2000 metres altitude above mean sea level. The downward movement of snowline in glacial age clearly indicates the lowering of temperature, particularly of the summer temperature. There is a difference of 1700 metres between the altitudes of the calculated glacial age snowline and the present snowline. By applying the standard lapse rate of 0.6°C/100 metres the above altitude difference can be converted into temperature equivalent.[19]

The glacial temperature was as much as 10.2°C lower than that of the present. The appreciable decrease in temperature during the glacier period was more prominent in summer when the temperature fell to nearly 3°C. Due to this change a larger amount of energy was spent on warming the ice and on sublimation than on melting.[20]

DISTRIBUTION OF GLACIERS

The glaciers in the Liddar valley are presently confined along the northern ridges of the East and the West Liddar valleys. The glaciers extend from two icefences on the north and the east from each of the two valleys.

West Liddar Valley Glaciers or Kolahoi Group of Glaciers

Presently there are six glaciers in the West Liddar valley. They are concentrated near Kolahoi mountain and cover an area of 20 sq km (Fig. 15). This glacier group is named after Kolahoi which is the largest glacier of the region. The following glaciers were investigated in this group.

Kolahoi Glacier

The Kolahoi glacier is the largest in the study region. It is 6.5 kilometres long and 300 metres wide at the snout located at an altitude of 3690 metres. The glacier headwall is located at an altitude of 5425 metres in Kolahoi mountain, located between two peaks, Dudnag (4952 metres above sea level) in the west and Hiurbagwan (4889 metres above sea level) in the east (Figs. 12, 15 and 17A). The glacier proper begins, at an altitude of 5000 metres, from Kolahoi mountain and descends in deep trough in two branches. The right branch is irregular and shrinking and the left massive and subdivides into two lobes by a pronounced central depression. Both the lobes merge into each other near an icefall at an altitude of 4000 metres. The glacier has northwest trend line and its surface is slashed by crevasses along the eastern margins. There are small caves along the eastern margin of the snout, which testify to large-scale meltwater production. Along the western margin there is an extensive mass of debris with ogives and crevasses. There are two meltwater outlets, the most easterly having

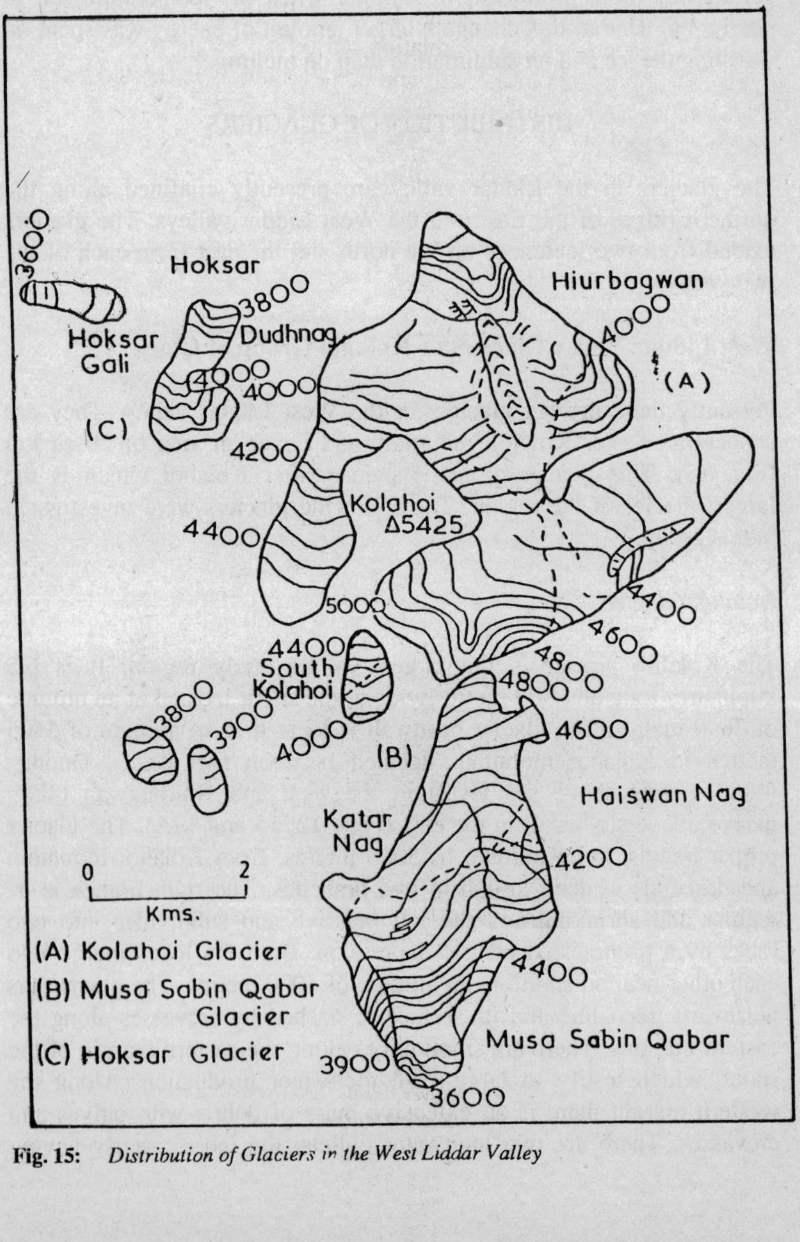

Fig. 15: *Distribution of Glaciers in the West Liddar Valley*

recently been developed in the glacier. These meltwater channels have their origin in the central depression located at an altitude of 4200 metres. The snout is flanked by a series of fresh moraines which mark its northern extension in early times. The southern limit is marked by glacial horn at Kolahoi peak (Fig. 16A). The firn limit to which the winter snow recedes during the ablation season has been observed at an altitude of 4500 metres in the east and at 4400 metres in the west. It is somewhat irregular. During the accumulation season the glacier advances to an altitude of 3400 metres. This line coincides with deposition of annual detrital material in the adjacent filled-up lake. By 1984 the glacier had retreated to an altitude of 3690 metres and by a distance of 210 metres. The snout has narrowed and moved away from lateral low moraines by a similar distance leaving behind end and lateral moraines near the front (Fig. 17A).

Fig. 16A: *Kolahoi Glacier Snout Slashed by Crevasses (A) Along the Eastern Margin, Extensive Mass of Ogives Basalt Dust (B) on the Western Margin and Meltwater Outlet (C) on Eastern Margin*

Fig. 16B: *Shishram Glacier Snout and its Icefield near the Confluence of Harnag Glacier (A) and Shishram Glacier (B) the Snout is Marked by Laminated Ice (C) and Ice-caves(D).*

Musa Sabin Qabar

It is the second largest glacier of the West Liddar, covering an area of 4.5 square kilometres, flowing southwestward. The glacier originates at an altitude of 4800 metres south of Kolahoi and terminates at an altitude of 3950 metres above sea level. The glacier headwall is located between the peaks of Katarnag mountain (4830 metres) in the west and Haiswaunar mountain (4400 metres) in the east (Fig. 15B). It is 1.8 kilometres long. Near the snout there are three well preserved terminal (end) moraines. The firn limit of the glacier ranges between the altitudes of 4600 and 4500 metres (Fig. 15B).

Hoksar

The glacier originates from Hoksar at an altitude of 4200 metres and terminates at 3600 metres. It is 1.2 kilometre long and has an area of 2.4 sq. km. It flows between Dudnag mountain in the west and the Hoksar in the east and has northwest orientation (Fig. 15C). The glacier is slashed by crevasses both longitudinally and transversely. The snout is flanked by low lateral moraines. These moraines are 1/2 kilometre long. The Hoksar glacier has deposited a huge end moraine of angular blocks in front of the snout, most of whose ground mass has been washed away. The firn limit of the glacier is located at an altitude of 4050 metres (Fig. 15C).

East Liddar Valley Glaciers or Shishram Group of Glaciers

There are eleven glaciers in the East Liddar valley, the largest being the Shishram glacier (Fig. 18). They cover an area of 30 sq. km. Most glaciers have their nevefield in the eastern icefence. The East Liddar glaciers are also called Shishram group of glaciers, named after the largest glacier. Some of the important glaciers observed in the field are discussed below.

Shishram Glacier

The glacier originates in Kunyirhayan cirque at an altitude of 4900 metres. The main icefield is on the eastern side of the peak and the glacier descends from it by an icefall to a large basin. There it meets the tongues of two glaciers originating from smaller icefields. From this basin the glacier descends towards the northwest and joins the Harnag glacier near Zamtirnag. The Shishram glacier is 5 kilometres long and 230 metres wide near Zamtirnag terminus. The terminus is located at an altitude of 3740 metres. The glacier is characterised by the presence of sharply serrated ridge along its southern and eastern side and prominent glacial horn at Kunyirhayan peak (Fig. 17B). The glacier has lobate form and protrudes from northwestern side of Zamtirnag Cirque at an angle of 45°. Near the snout the ice reveals characteristic laminated structure from which individual laminae move out differentially (Fig. 16B). The snout is flanked by lateral moraines on either side, but one to the west is more massive and better preserved. The tip of the snout on the western side is covered by recent 8 metres high end moraine extending for about 275 metres. The other two end moraines are observed near Zamtirnag mouthing.

Loosely consolidated ice has been observed just near the central notch at an average height of 4400 metres where ice has an average density of 0.51 gm/cm^3 and both accumulation and ablation measurements in the glacier were somewhat uniform. During the 1984 winter season the snowline advanced to 3400 metres as glacier moved towards the Shishramnag lake, where icebergs float in early summer (Fig. 16B). From an altitude of 3550 metres in 1923 the glacier has retreated by 190 metres to an altitude of 3740 metres in 1984.

Sonsar-Natsar Glacier

Sonsar-Natsar is a small valley glacier. It originates at 4700 metres and terminates at an altitude of 3900 metres above mean sea level. In the absence of any name given to it by the Survey of India, it has been named after the channel which originates from it. The glacier is 2 kilometres long and 230 metres wide at the snout. It flows in the northwest direction. The glacier has its neve field near Kaltip mountain cirque at an altitude of 4700 metres. At the back of the cirque there is prominent glacial horn of Kaltip peak at an altitude of 5100 metres (Fig. 18B). It marks the southern limit of the glacier.

The glacier has a firn line at 4400 metres where Natsar channel originates.

Dudhal Glacier

The small cirque glacier of Dudhal originates from two icefields located at an altitude of 4800 metres above mean sea level in Dudhal mountains. It terminates at an altitude of 3800 metres above mean sea level to the northwest, covering a distance of 2.3 kilometres. The glacier has a sinuous course and flows between Dudhal mountain in the west and Gillinkan mountain in the east. It has deposited a huge end moraine at 2700 metres above mean sea level near Chandanwari (Fig. 18C).

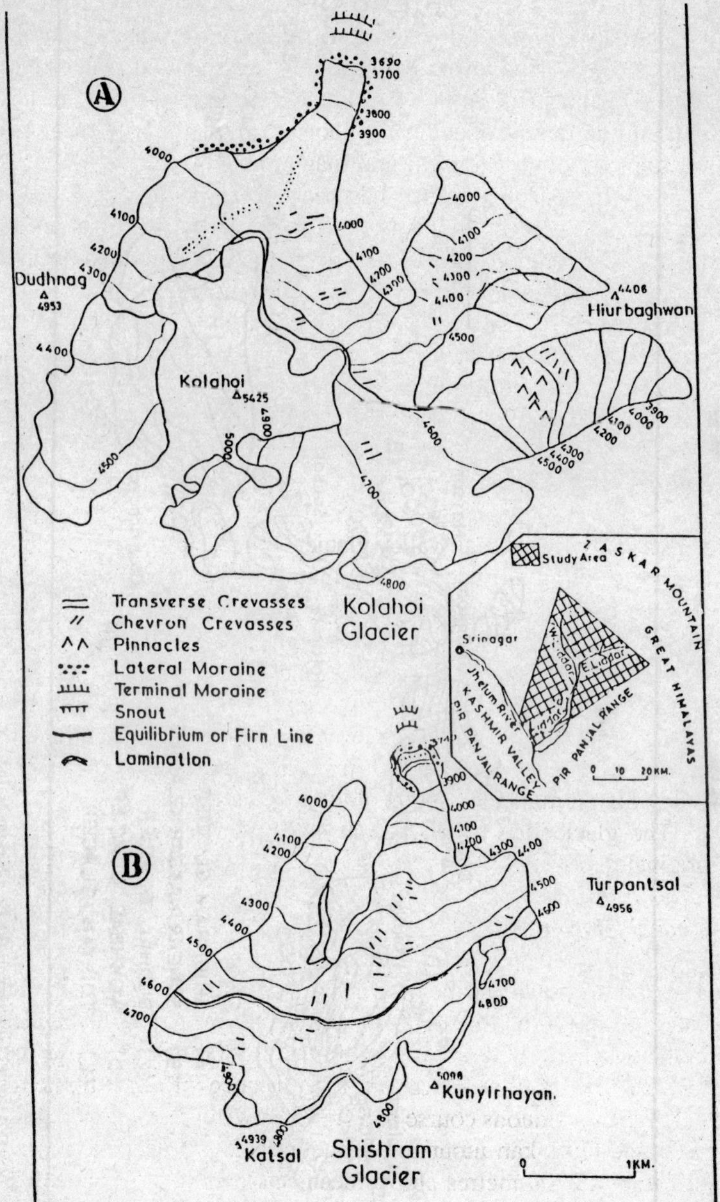

Fig. 17: *Kolahoi and Shishram Glaciers*

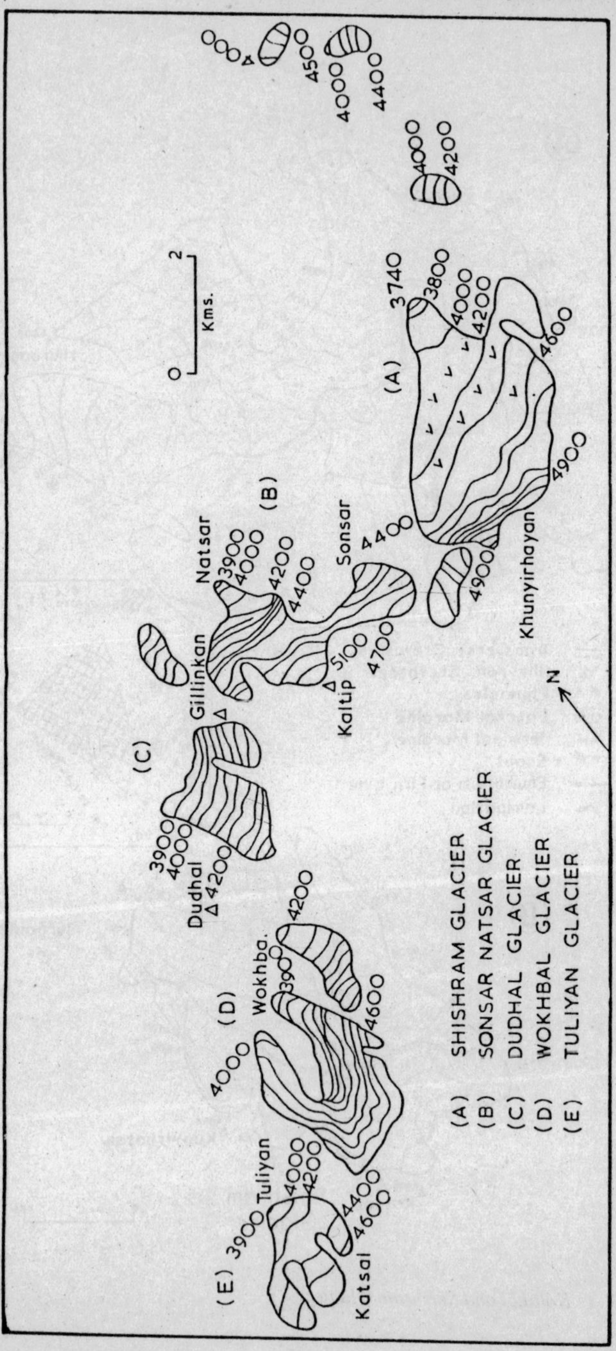

Fig. 18: *Distribution of Glaciers in the East Liddar Valley*

Tuliyan Glacier

Tuliyan is a small valley glacier, originating from three cirques located at an altitude of 4600 metres. They coalesce at an altitude of 4200 metres, from where it moves like a tongue along Tuliyan mountain. The glacier terminates at an altitude of 3700 metres. It has deposited a huge ground moraine with drumlins to the east of Pahalgam, near the valley bed (Fig. 18E).

MASS BALANCE OF THE GLACIERS

Mass balance is the most meaningful description of hydrological cycle of glacier and is the best means of estimating the glacier movement and in understanding the glacial process. It has been the most extensively investigated problem in the study of glacial geomorphology of Alps in Europe and of Rockies in North America.[21, 22, 23, 24, 25, 26] In contrast only a few studies on the mass balance of Himalayan glaciers, particularly in Kashmir Himalaya, have been conducted.[27, 28, 29] In the study region no mass balance measurements have been made. The present investigation is a preliminary contribution.

Concept of Mass Balance

Mass balance is a theoretical state of glacier in which accumulation of snow and loss of ice mass through ablation tends to be equal. This state is rarely if ever attained. Mass balance in the life history of glacier at any particular time stage is either negative or positive, displaying a deficit or surplus of ice. These two conditions are the principal determinants of glacier movement. While deficit results in retreat, surplus results in advance. Advance and retreat of glacier in their own turn generate glacial processes which create the erosional and depositional features in glacial geomorphological landscape (Fig. 19). The processes are related to the movement and ultimately mass balance is generated by specific climatic conditions.[30]

Elements of Mass Balance

Mass balance of glacier is mainly related to temperature and precipitation, but other micro-meteorological elements, such as radiation, humidity, evaporation and wind direction play their secondary role.

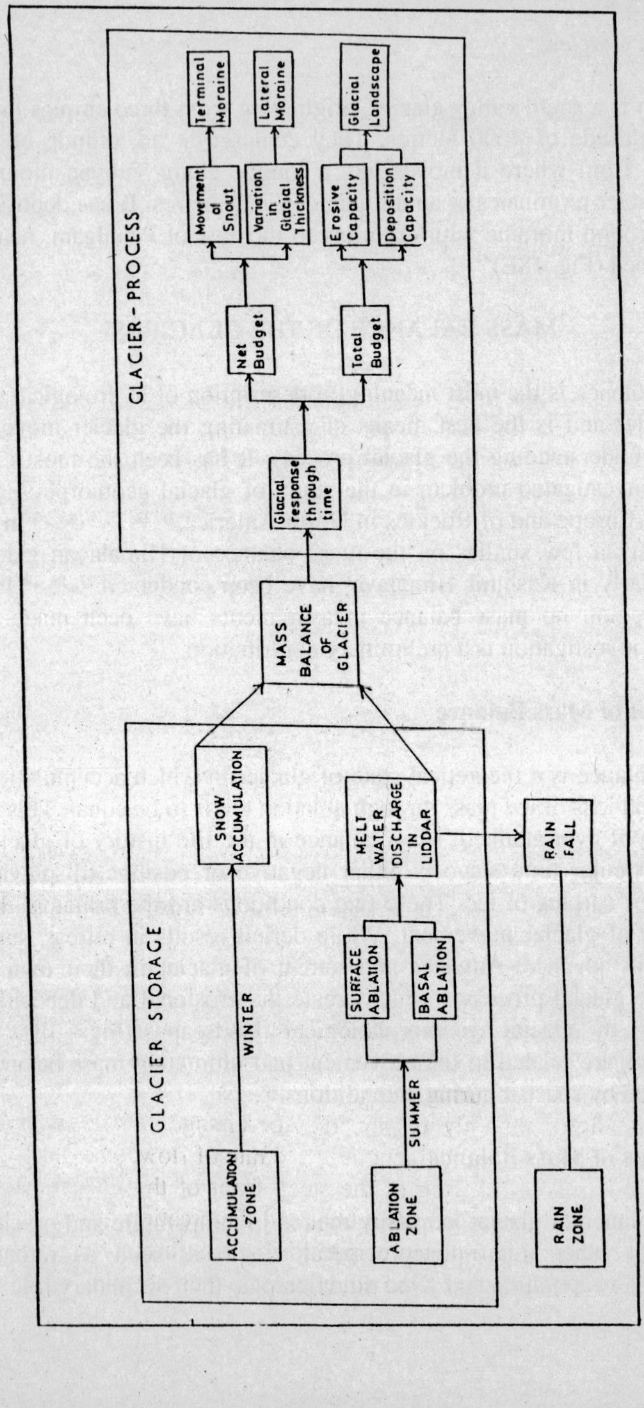

Fig. 19: *The System of Mass Balance*

Temperature Correlate

Temperature is an important element of mass balance. Its variations effect the surface and basal melting of glacier and it reasonably well correlates with ablation. The average summer temperature gives rough estimate about net annual ablation. Schytt estimated that 1°C temperature corresponds to 41 centimetres of ablation.[31] This relationship correlates closely with the field measurements in the study region. The Liddar glaciers have an average summer temperature of 18.8°C, which corresponds to 770.8 centimetres (7.71 metres) of ablation (Table 5).

Table 5: *Mean Monthly Temperature and Rainfall of Chandanwari, 1983*

Months	*Mean Maximum (°C)*	*Mean Minimum (°C)*	*Rainfall (millimetres)*
January	2.0	–7.0	167.5
February	4.0	–10.0	112.5
March	15.0	0.0	184.0
April	18.5	2.0	122.5
May	24.0	4.5	133.5
June	24.5	5.0	220.0
July	23.5	7.0	129.0
August	22.0	7.5	76.0
September	19.0	8.0	24.0
October	11.0	–5.0	8.5
November	8.0	–9.5	5.5
December	0.0	–14.5	185.0
Annual			1370.0

Source: Indian Meteorological Department, Srinagar, *Monthly Weather Reports*, 1983.

Precipitation Correlate

Precipitation has a strong control over mass balance as it brings positive change in net balance of glacier due to snow accumulation in catchment areas during the winter season. The annual change in accumulation is mainly responsible for change in the snout position, thickness of glacier, and length of the rate of flow. The positive budget results in the advancing of the steep front of the glacier whereas the negative budget induces its retreat and the creation of gentle slopes. In the summer season the precipitation in the form of rainfall near the snout of the glaciers of the study region does not play any part in accumulation as it helps in the surface melting of the glacier near the snout due to liberation of heat by latent heat of condensation. The snow and ice melt contribute nearly 92 per cent of the total discharge of the Liddar river. The high percentage of meltwater is due to high radiation melt, accounting for 80 to 90 per cent of the total local balance. The maximum radiation balance recorded in the Himalaya was 28 mwm^2 per day in August 1974.[32]

Methods of Analysis and Results

The mass and total balance of the Liddar glaciers were measured by the following methods:

(1) The mass balance of glacier was measured by input-output relation of snow and ice with respect to net accumulation and net ablation. The net accumulation or winter balance was measured in terms of water equivalent volume in April 1984. The net ablation or summer balance was measured in terms of water equivalent volume during the summers of 1983 and 1984. This value was subtracted from the net accumulation of 1984 to compute the mass balance.

(2) By applying suitable climate-glacier model. The input parameters of such model were precipitation, discharge rate and temperature, whose data were created through field observations and climatic record.

(3) The total balance of the present and past glaciers and their fluctuations were evaluated from historical records, moraine morphology and vegetation succession.

Accumulation or Winter Balance

The field measurements of the total winter accumulation were carried out at the end of the accumulation season during the month of April 1984 along Kolahoi and Shishram glaciers in the Eastern and the Western parts of the Liddar valley. The most common technique adopted was digging pits at different elevations both in glacier as well as in the adjoining areas, combined with network of snowdepth sounding to assess the accumulation, density and water equivalent of snow.[33, 34, 35] The snowdepth was measured by taking sounding with 4 metre long metallic rod with pointed ends through new snow to the glacier ice surface (Fig. 20A).

The sounding measurements of snow were taken for Pahalgam, Laripora, Chandanwari, Aru and Liddarwat, because no snowfall data was available either from meteorological department or from any other State Government department. The snowpits were excavated with shovel at some of the sounding heights of glaciers to confirm their total accumulation, density and water equivalent (Fig. 20B). Forty snow pits were dug, 20 each in Kolahoi and Shishram glaciers. The different snow samples were collected from the pits by 40 centimetres long snow sampler (Fig. 21). The samples so collected were weighed in spring balance. The densities of samples were calculated by dividing the weights of samples by known volume of sampler. Then water equivalents were computed by multiplying the densities with the length of the sampler. The accumulation of winter snow in Kolahoi and Shishram glaciers ranged between 1 metre and 3.2 metres and 1 metre and 4 metres respectively (Tables 6 and 7). The densities and water equivalent of snow also varied with accumulation. The density increased sharply from 0.41 to 0.60 gram per cubic centimetre up to the depth of 220 centimetres. Below that it remained constant. This

Fig. 20A: *Measurement of Depth of Accumulated Snow of Kolahoi Glacier by Sounding Method*

Fig. 20B: *Excavation of Snowpit by Shovel in Accumulation Zone of the Kolahoi Glacier to find out Net Accumulation. The Pit is Dug to the Previous Summer Surface Marking of Ice*

variation was noticed in higher altitudes of two glaciers around 4600 metres. It may be due to fresh snow cover. In the higher altitude above 4600 metres the avalanches disturbed the stratification and made the traverse very difficult.

The volume of ice and its water equivalent were computed for accumulation and ablation zone of the glaciers after every 100 metres contour interval. The value ranges between 50 centimetres and 2.0 metres of water equivalent (Fig. 23) (Table 10). The total winter accumulation volume in Kolahoi and Shishram glaciers amounted to 17.39 X 10^6 m^3 and 14.2 X 10^6 m^3 respectively. Their accumulation rate in water equivalent amounted to 1.464 X 10^6 m^3/km^2 (146.4 gm/cm^2) and 1.430 X 10^6 m^3/km^2 (143 gm/cm^2) respectively. The difference in the rate of accumulation in the two glaciers seems to be a variation in accumulation areas over the entire length of the

Fig. 21A: *Measuring the Volume of the Snowpit by 40 Centimetre Long Snow Sampler. The Snow is Collected in the Sampler by Pressing it into the Snow*

glacier. The upper trasverse profile above the equilibrium line at 4500 metres in two glaciers suggest that the Kolahoi glacier has lesser area in accumulation zone than the Shishram glacier, which reflects large accumulation in the latter.

Ablation

The loss of ice and snow from the glaciers resulting from their melting during the summer season is termed ablation. Ablation is a function of air temperature, hence, it varies with altitude. It is larger in the lower and smaller in the higher altitudes and lowest near the highest altitude of the glacier. It is of two types: (i) surface and (ii) basal.

Fig. 21B: *Finding out the Density and Water Equivalent of Snowpack of the Shishram Glacier — the Known Volume of Snow is Weighed by Spring Balance in the Field*

Table 6: *Snowpit Measurements in the Accumulation Zone of the Shishram Glacier, 1984*

Pit	*Number of samples*	*Length of snow sampler (cm)*	*Length of sample snow column (cm)*	*Weight of the snow sample column (cm)*	*Density of snow sample (gm/cm³)*	*Water equivalent of snow sample (gm/cm²)*	*Cumulative water equivalent of snow pit samples (gm/cm²)*
1	2	3	4	5	6	7	8
1	1	40	40	720	0.41	16.4	16.4
	2	40	30	700	0.46	13.8	30.2
2	1	40	40	750	0.43	17.2	17.2
	2	40	40	820	0.47	18.8	36.0
	3	40	40	860	0.48	19.2	55.2
	4	40	40	930	0.52	20.8	76.0
	5	40	20	450	0.51	10.2	86.2

(Contd.)

Table *(Contd.)*

1	2	3	4	5	6	7	8
3	1	40	40	705	0.42	16.8	16.8
	2	40	40	720	0.41	16.4	33.2
	3	40	40	960	0.55	22.0	55.2
	4	40	40	900	0.51	20.4	75.6
	5	40	40	930	0.53	20.0	95.6
	6	40	40	940	0.53	20.0	115.6
	7	40	16	380	0.54	8.6	124.2
4	1	40	40	825	0.47	18.8	18.8
	2	40	40	860	0.48	19.2	38.0
	3	40	40	900	0.54	21.6	59.6
	4	40	40	940	0.53	20.0	79.6
	5	40	40	945	0.54	21.6	101.2
	6	40	40	940	0.53	20.0	121.2
	7	40	40	980	0.56	22.4	143.6
	8	40	40	985	0.56	22.4	166.0
	9	40	40	985	0.56	22.4	188.4
	10	40	40	980	0.58	23.6	212.0
	11	40	20	195	0.60	18.6	230.6

Source: Field Measurements.

Table 7: *Snowpit Measurements in the Accumulation Zone of the Kolahoi Glacier, 1984*

Pit	*Number of samples*	*Length of snow sampler (cm)*	*Length of sample snow column (cm)*	*Weight of the snow sample column (cm)*	*Density of snow sample (gm/cm^3)*	*Water equivalent of snow sample (gm/cm^3)*	*Cumulative water equivalent of snow pit samples (gm/cm^2)*
1	2	3	4	5	6	7	8
1	1	40	40	760	0.43	17.2	17.2
	2	40	40	820	0.46	18.4	35.6
	3	40	40	990	0.56	22.4	58.0
	4	40	40	905	0.51	20.4	78.4
	5	40	10	125	0.52	2.6	81.0
2	1	40	40	735	0.42	16.8	16.8
	2	40	40	840	0.48	19.2	36.0
	3	40	40	925	0.52	20.8	56.8
	4	40	40	900	0.51	20.4	77.2
	5	40	40	960	0.55	22.0	99.2
	6	40	30	740	0.56	16.8	116.0
3	1	40	40	845	0.48	19.2	19.2
	2	40	40	872	0.50	20.0	39.2
	3	40	40	905	0.52	20.8	60.0
	4	40	40	955	0.54	21.6	81.6

Contd.)

Table (*Contd.*)

1	2	3	4	5	6	7	8
	5	40	40	955	0.54	21.6	103.2
	6	40	40	980	0.56	22.4	125.6
	7	40	40	985	0.56	22.4	148.0
	8	40	16	405	0.58	9.3	157.3

Source: Field Measurements.

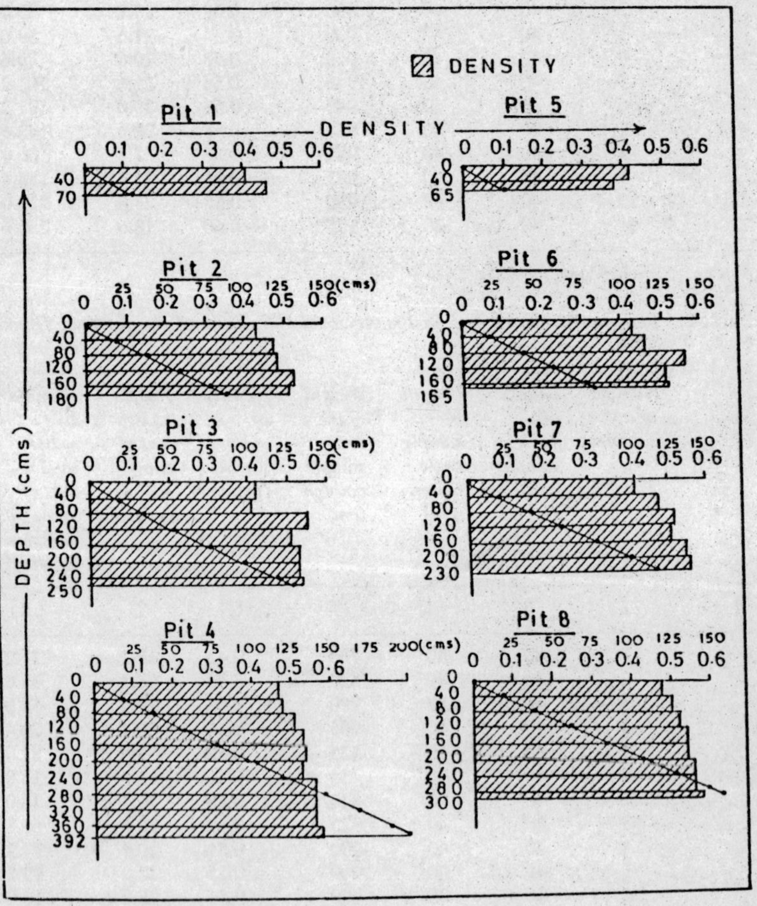

Fig. 22: *Depth-Density Relationships in Snowpits*

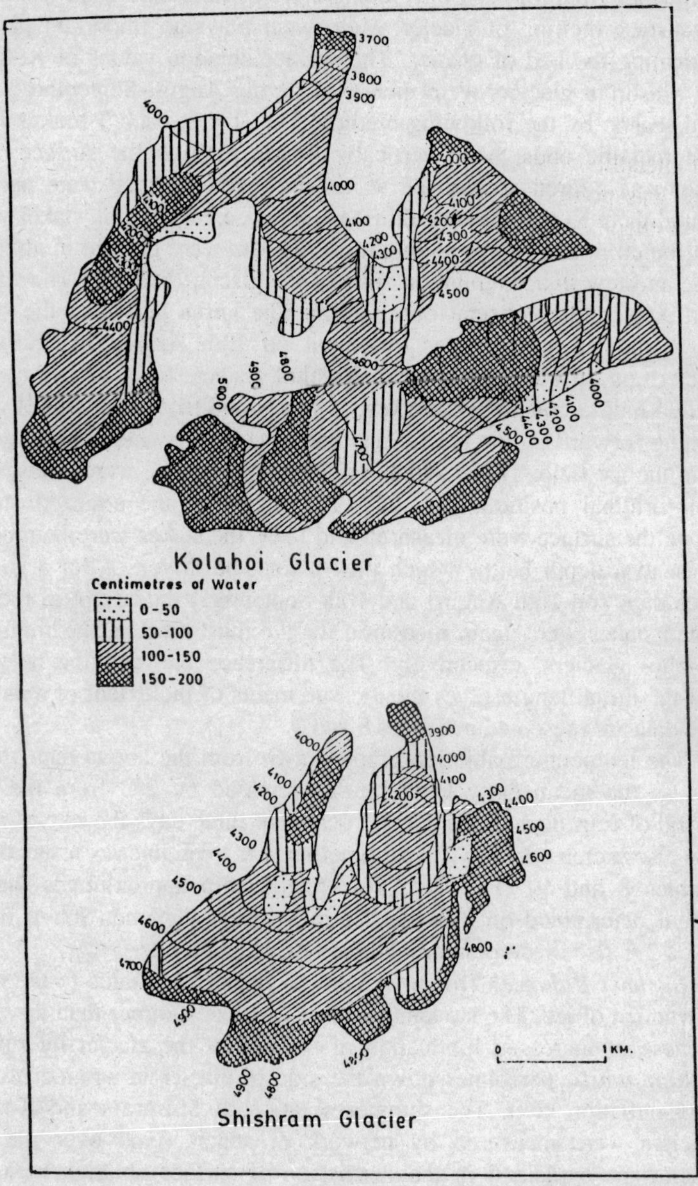

Fig. 23: *Liddar Valley Glaciers: Winter Accumulation in Water Equivalent (1983-84)*

Surface Ablation: High air temperature of summer season results in the surface melting of glacier. Melt water plays an important part in lubricating the bed of glacier. The surface ablation values of Kolahoi and Shishram glaciers were measured during August-September 1983 field season by the following method. Ten stakes, each 3 metres long with metallic ends, were vertically driven through the surface. Hot water was poured around the stakes and snow and ice were packed around them so that they would freeze in place.[36] These ten stakes were distributed in a 'S' form so that most of them were planted in ablation zone and few in accumulation zone of the glacier (Fig. 24). The stakes were kept under constant observation. The initial height of the stake above the ice surface was measured on 10th August in Shishram glacier and on 4th September in Kolahoi glacier. After three days the second measurements were taken. It was found that many stakes were leaning forward on the ice surface due to constant wind coming down from the ice field. Those which were leaning forward were adjusted to their original position (Fig. 25). The height of the adjusted stakes above the surface were measured and later, the stakes were hammered down to a depth below which they cannot be driven. After a gap of three days (on 16th August and 10th September) the height of the adjusted stakes were again measured for the third time in Shishram and Kolahoi glaciers, respectively. The difference between the adjusted and the initial length gives the measurements of the extent of ablation. The data are presented in Tables 8 and 9.

The temperature-ablation graph drawn from the above data shows that as the mean daily temperature increased by 3°C from the first period of four days to the second period of three days the rate of ablation also increased from 3.1 centimetres to 6.4 centimetres respectively (Tables 8 and 9) (Fig. 26). This relationship approximates that of Schytt determined on Stos glacier in Swedish Lapland. Schytt found that a 2°C rise in temperature doubled the ablation rate.[37]

Summer Balance: The movement of glaciers is related to the basal movement of ice. The ice tends to flow faster in summer than in winter because of increased lubrication of the bed of the glacier by surface melting which percolates down the side of glacier in summer due to high radiation melt. The summer ablation of Shishram and Kolahoi glaciers were measured by network of stakes fixed over the two glaciers as explained in the discussion of surface ablation measurements and also by observing the pits already dug during the April field season. Some of the new pits were dug into accumulation and ablation zones of the glaciers during August-September 1984 field season. In ablation zone blocks of ice were excavated by ice axe at different sites to measure their densities and to compute their water equivalent. The volume of ablation was computed for each glacier after every 100 metres of height. The gross ablation in Kolahoi and Shishram glaciers varied between 20.54 X 10^6 m^3 and 17.05 X 10^6 m^3 equivalent of water respectively. Their ablation rate also varied between 1.73 X 10^6 m^3 / km^2 (173 gm/cm^2) and 1.715 X 10^6 m^3 / km^2 (171.5 gm/cm^2) respectively.

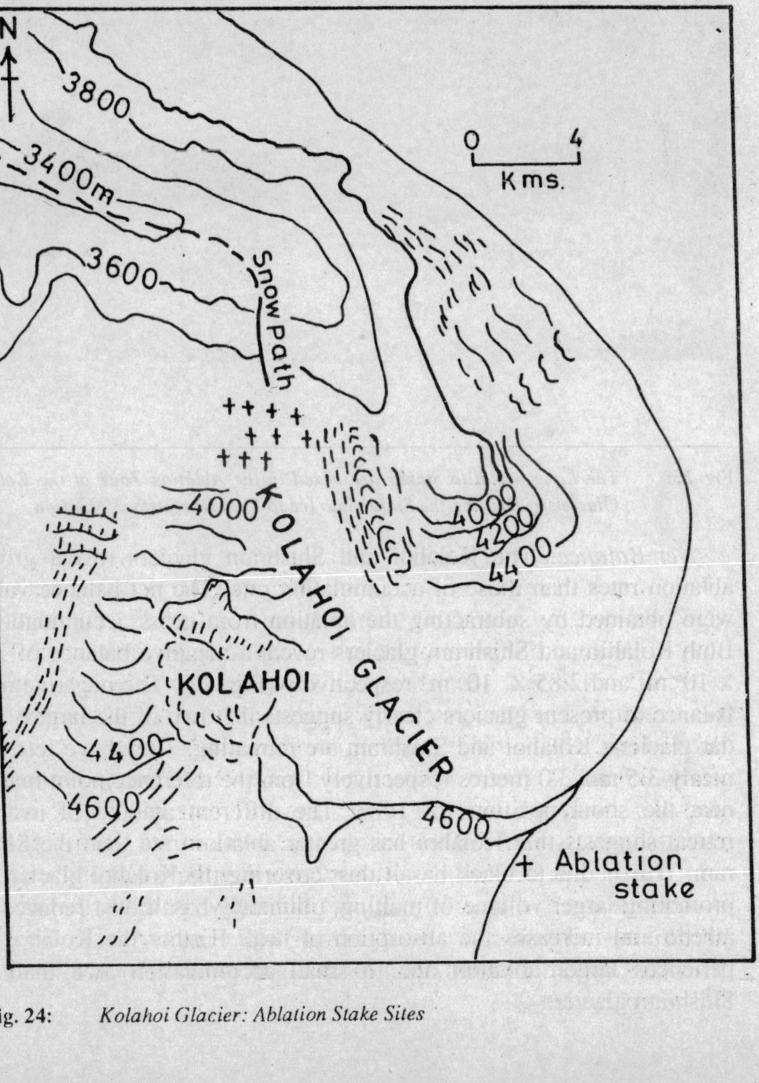

Fig. 24: *Kolahoi Glacier: Ablation Stake Sites*

Fig. 25: *The Long Metallic Stakes are Fixed in the Ablation Zone of the Kolahoi Glacier by Drilling the Stakes into Ice to Find Out Surface Ablation*

Net Balance: The Kolahoi and Shishram glaciers reveal greater ablation rates than those of accumulation rate. The net balance values were obtained by subtracting the ablation from gross accumulation.[38] Both Kolahio and Shishram glaciers reveal a negative balance of 3.15 x 10^6 m^3 and 2.85 x 10^6 m^3 respectively (Fig. 27). The negative mass balance of present glaciers clearly suggests that two of the largest Liddar glaciers; Kolahoi and Shishram are retreating. They have retreated nearly 3.5 and 3.0 metres respectively from the reference point marked near the snout positions of 1983. The different amount of average retreat suggests that Kolahoi has greater ablation rate than the Shishram. This is due to black basalt dust covering the Kolahoi glacier and promoting larger volume of melting, ultimately basalt dust reduces the albedo and increases the absorption of heat. Further the Kolahoi experiences larger ablation due to small accumulation area than the Shishram glacier.

Hydrological Model

The changes in the mass balance of the Liddar glaciers in general and the East and the West Liddar glaciers in particular were evaluated by hydrological model.[39, 40, 41] The input parameters of this model were precipitation and average discharge rate of meltwater channel. The model was applied to the Aru (West Liddar), the Chandanwari and the Shishnag (East Liddar) group of glaciers, because their meltwater directly enters through braided streams into each of their drainage basins. The annual discharge at each station reflects the amount of glaciers melting in the West and the East Liddar, respectively (Table 11). The model was applied to the Shishram glacier to corroborate the result obtained as discussed earlier. Its meltwater directly enters the Shishnag lake through Zamtirnag stream which emerges at the terminus of the glacier. The gauging station is located at the mouth of the East Liddar, where the Shishnag drains into the East Liddar.

Table 8: *Surface Ablation Estimates of Shishram Glacier, 1983*

Sl. No.	*Stakes*	*Ablation Stake Measurements*				*Ablation Rate (cm)*
		Tenth August	*Thirteenth August*		*Sixteenth August*	
		Initial Length of Stake (cm)	*Adjusted Length of Stake (cm)*	*Initial Length of Stake (cm)*	*Adjusted Length of Stake (cm)*	
1	4	138.6	145.4	148.2	170.4	29.0 (10 - 16 August)
2	5	140.5	151.7	151.7	172.6	32.1 (10 - 16 August)
3	2	152.8	162.4	163.7	Down	9.6 (10 - 13 August)
4	1	128.9	140.2	138.2	160.2	33.3 (10 - 16 August)
5	8	135.0	Down	138.4	150.8	12.4 (13 - 16 August)
6	7	Metal	Stake	Lost		
7	3	142.6	152.8	154.2	Down	10.2 (10 - 13 August)
8	6	136.4	154.0	149.6	164.8	32.8 (10 - 16 August)
9	10	Metal	Stake	Lost		
10	9	132.6	Down	137.8	146.6	8.8 (13 - 16 August)

Source: Field Measurements.

Table 9: *Surface Ablation Estimates of Kolahoi Glacier, 1983*

Sl. No.	*Stakes*	*Ablation Stake Measurements*				*Ablation Rate (cm)*
		Fourth September	*Seventh September*		*Tenth September*	
		Initial Length of Stake (cm)	*Adjusted Length of Stake (cm)*	*Initial Length of Stake (cm)*	*Adjusted Length of Stake (cm)*	
1	2	143.7	156.7	Down	Down	13.0 (4 - 7 September)
2	9	132.8	142.8	140.5	170.0	39.5 (4 - 10 September)
3	8	125.4	133.7	131.8	158.3	34.8 (4 - 10 September)
4	6	153.0	170.4	Down	Down	16[illegible] (4 - 7 September)
5	4	Metal	Stake	Lost		
6	3	140.6	151.2	148.6	179.2	41.2 (4 - 10 September)
7	7	137.2	Down	140.2	148.3	8.1 (7 - 10 (September)
8	1	119.7	134.1	133.4	Down	14.4 (4 - 7 September)
9	5	122.9	135.6	145.1	169.6	37.2 (4 - 10 September)
10	10	96.6	106.5	107.8	Down	9.9 (4 - 7 September)

Source: Field Observations.

Average Ablation rate during:

(i) 4th September to 7th September = 3.1 cm

(ii) 4th September to 10th September = 6.4 cm

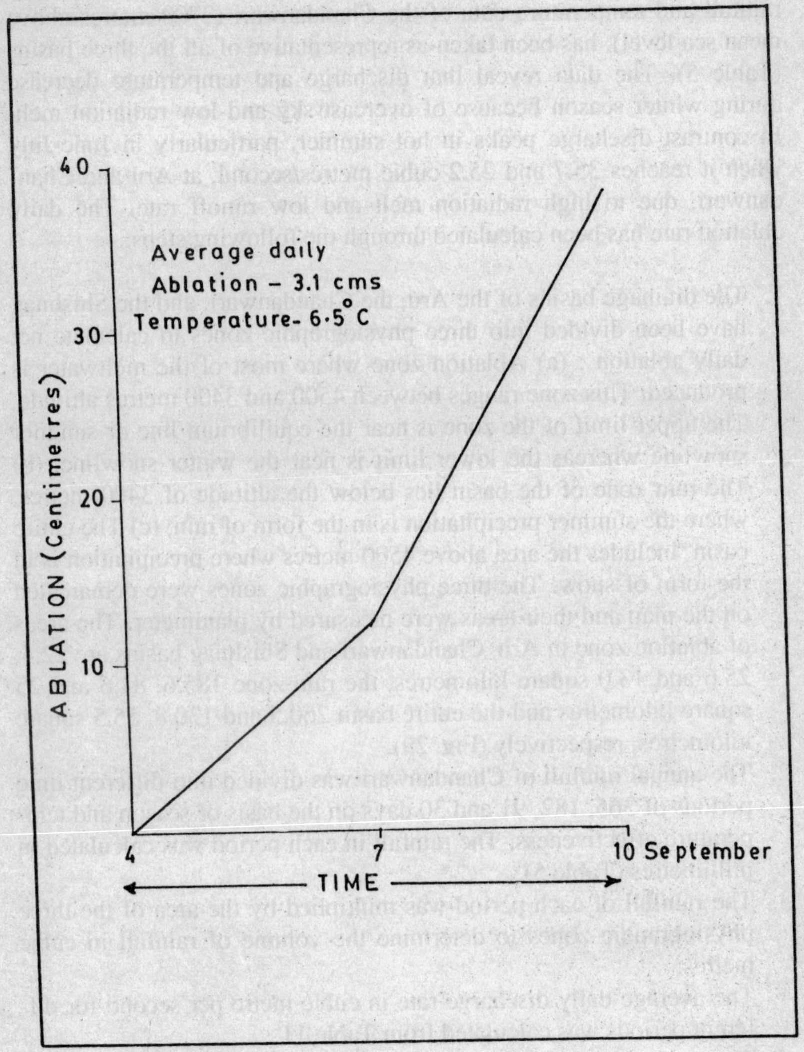

Fig. 26: *Kolahoi Glacier : Temperature-Ablation Relationship*

The discharge at each of the gauging station reflects the amount of glacier melting and precipitation run-off. The amount of discharge reflects the result of temperature condition prevailing in the glacier.

Since there is no rain-gauge station near the Shishnag and the Aru, the rainfall and temperature data of the Chandanwari (2700 metres above mean sea level), has been taken as representative of all the three basins (Table 5). The data reveal that discharge and temperature decrease during winter season because of overcast sky and low radiation melt. In contrast discharge peaks in hot summer, particularly in June-July when it reaches 35.7 and 25.2 cubic metres/second, at Aru and Chandanwari, due to high radiation melt and low runoff rate. The daily ablation rate has been calculated through the following steps.

1. The drainage basins of the Aru, the Chandanwari, and the Shishnag have been divided into three physiographic zones to calculate net daily ablation : (a) Ablation zone where most of the meltwater is produced. This zone ranges between 4500 and 3400 metres altitude. The upper limit of the zone is near the equilibrium line or summer snowline whereas the lower limit is near the winter snowline, (b) The rain zone of the basin lies below the altitude of 3400 metres, where the summer precipitation is in the form of rain, (c) The entire basin includes the area above 4500 metres where precipitation is in the form of snow. The three physiographic zones were demarcated on the map and their areas were measured by planimeter. The areas of ablation zone in Aru, Chandanwari and Shishnag basins are 22.4, 25.6 and 13.0 square kilometres, the rain zone 185.6, 84.6 and 25 square kilometres and the entire basin 260.5 and 130.8, 55.5 square kilometres, respectively (Fig. 29).
2. The annual rainfall of Chandanwari was divided into different time periods of 365, 182, 91 and 30 days on the basis of season and temperature effectiveness. The rainfall in each period was calculated in millimetres (Table 5).
3. The rainfall of each period was multiplied by the area of the three physiographic zones to determine the volume of rainfall in cubic metres.
4. The average daily discharge rate in cubic metre per second for different periods was calculated from Table 11.
5. The total discharge of each time period was converted from cubic metre per second to cubic metre per day by multiplying the average daily discharge by 3600 x 24 and then by the number of days in each period.
6. The volume of meltwater flowing through each physiographic zone was calculated by subtracting the rainfall volume from the total discharge.

Table 10: *Kolahoi-Shishram Glacier Mass Balance Estimates (in Water Equivalent), 1983-84*

Elevation (metres)	*Kolahoi Glacier*				*Shishram Glacier*			
	Area (km^2)	*Accumulation (10^6m^3)*	*Ablation (10^6m^3)*	*Net Balance (10^6m^3)*	*Area (km^2)*	*Accumulation (10^6m^3)*	*Ablation (10^6m^3)*	*Net Balance (10^6m^3)*
(a) *Ablation Zone*								
3600 - 3700	0.107	0.194	0.228	–0.034	—	—	—	—
3700 - 3800	0.243	0.284	0.518	–0.234	0.248	0.446	0.616	–0.170
3800 - 3900	0.650	0.826	1.643	–0.817	0.467	0.520	1.102	–0.582
3900 - 4000	0.625	1.216	3.458	–2.242	0.465	0.365	0.969	–0.604
4000 - 4100	1.248	1.248	2.563	–1.315	0.625	0.888	1.487	–0.599
4100 - 4200	1.750	2.065	3.428	–1.363	0.630	0.883	1.496	–0.613
4200 - 4300	1.150	1.356	2.425	–1.069	0.816	1.192	1.950	–0.758
4300 - 4400	1.300	1.776	2.445	–0.669	0.925	1.554	3.192	–1.638
4400 - 4500	1.515	2.128	2.335	–0.207	1.755	2.204	4.282	–2.078
Total	8.588	11.093	19.043	–7.950	5.931	8.052	15.094	–7.042
(b) *Accumulation Zone*								
4500 - 4600	1.875	3.146	1.036	+2.110	1.375	1.717	0.884	+0.833
4600 - 4700	1.020	2.180	0.265	+1.915	0.925	1.650	0.698	+0.952
4700 - 4800	0.312	0.682	0.168	+0.514	1.155	1.752	0.301	+1.451
4800 - 4900	0.095	0.190	0.016	+0.174	0.525	1.036	0.072	+0.964
4900 - 5000	0.028	0.095	0.012	+0.083	—	—	—	—
Total	3.330	6.293	1.497	+4.796	3.980	6.155	1.955	+4.200
Total (a) + (b)	11.918	17.386	20.540	–3.154	9.911	14.207	17.049	–2.842

Source: Field Observations.

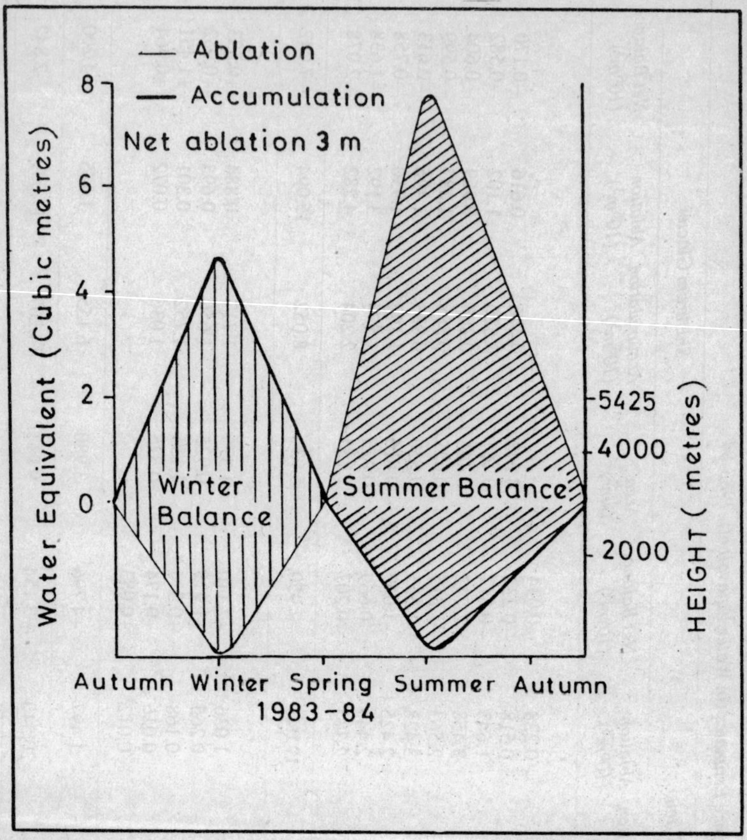

Fig. 27: *Mass Balance of Liddar Glacier*

7. The meltwater produced by ablation was computed by dividing the volume of meltwater flowing through each physiographic zone (as computed in the step 6) by the area of ablation zone (Tables 12, 13 and 14).
8. Each estimate of meltwater in three zones was divided by the number of days in each time period to estimate the ablation rate per day.[32]

Table 11: *Average Monthly Discharge at the Selected Sites of the East and the West Liddar Rivers. (m^3 /s), 1983-84*

Months	*Gauge Station*		
	Shishnag (East Liddar)	*Chandanwari (East Liddar)*	*Aru (West Liddar)*
January	5.2	5.7	6.2
February	4.4	5.2	5.9
March	5.8	7.3	6.9
April	7.0	14.3	20.3
May	14.2	20.6	35.7
June	16.8	25.2	34.2
July	14.3	21.6	26.4
August	14.4	19.2	21.4
September	8.2	9.6	14.8
October	4.5	5.5	9.9
November	3.8	5.3	9.2
December	3.7	4.4	8.5
Mean	8.5	12.0	16.6

Source: Power Development Department, Jammu & Kashmir Government, Srinagar.

The estimates for net balance apparently reflects the trend of fluctuations in net surface melting of ice/snow in water equivalent in three physiographic zones of the basin (Tables 12, 13 and 14, Step 7 above). The highest (21.96 metres) estimates are found in ablation zone of glacier and the lowest (7.5 metres) in the entire zone of the basin. The former is the source zone of meltwater production and latter the transfer zone. The total annual ablation in different physiographic zones of Kolahoi and Shishram glaciers ranged between 9.5 and 21.96 metres and 14.84 to 19.23 metres respectively (Tables 13 and 14). Nearly 80 per cent of this occurred during summer season and the rest in spring and winter seasons. The average daily ablation in the Kolahoi (West Liddar) and the Shishram (East Liddar) glaciers decreased from 9.4 centimetres/day and 7.9 centimetres/day to 2.5 cm/day between summer and winter season respectively. This suggests that more than two-thirds of the total ablation in Liddar glacier occurs between May and July (Tables 12, 13 and 14).

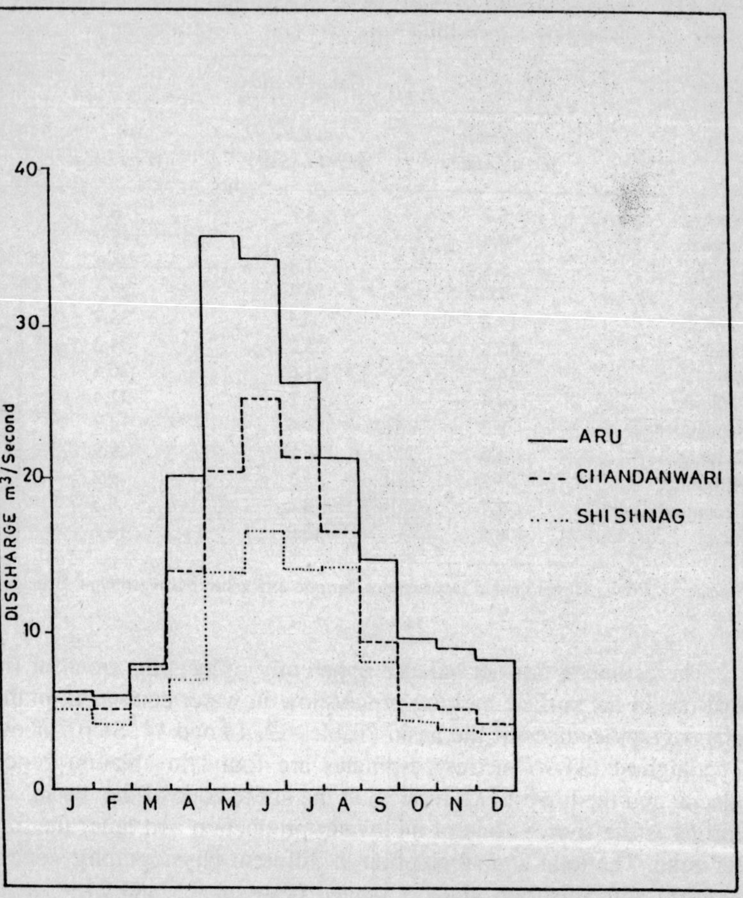

Fig. 28: *Monthly Average Discharge of Liddar River (Selected Stations) (1983)*

The studies mentioned above reveal that glaciers of the West Liddar valley are ablating at a faster rate than their counterpart in the East Liddar valley and glaciers have overall negative mass budget. The results of hydrological measurements closely correlate with those of the mass balance and surface ablation recorded in the field.

The average ablation rate for the year 1983-84 has ranged between 2 and 5 Centimetres per day, which correlates reasonably well with the surface ablation rate as well as with the total net ablation measurements observed in the field while calculating the mass balance of

Shishram glacier. It is highest in June-July when it peaks to 10 centimetres per day. In winter it falls to less than 0.1 centimetre per day.

FLUCTUATION RECORDS

Glacier fluctuations are indicated by the retreat and advance of the glacier snout over a period of few years or decades. It is caused by the

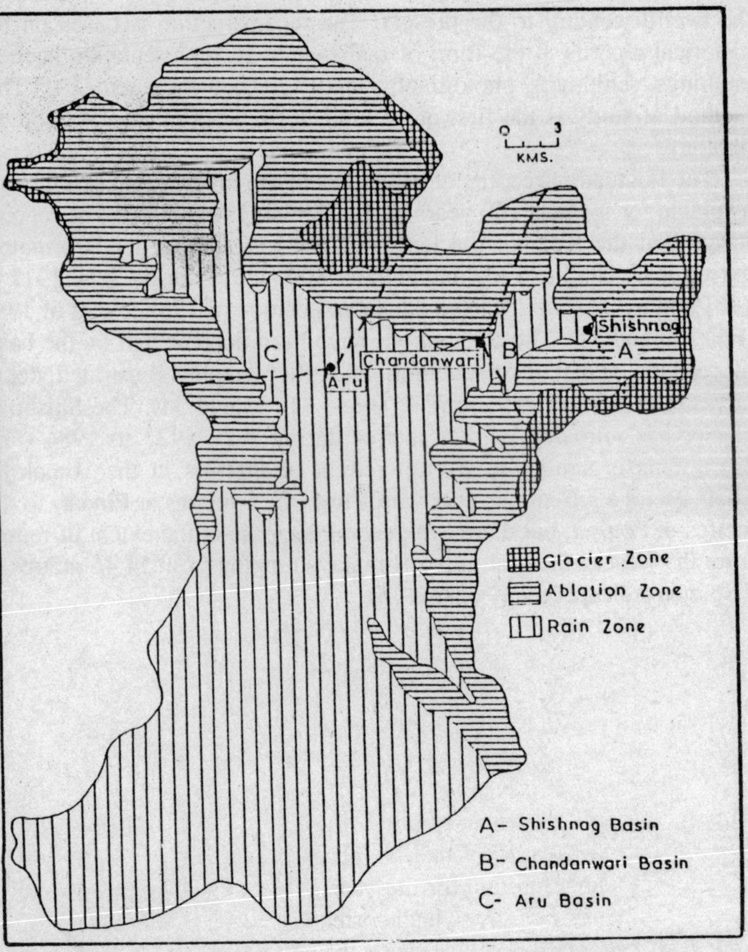

Fig. 29: *Liddar Valley : Physiographic Divisions*

changes in mass balance in response to climatic fluctuations. Mass balance records are available for short span of time. The fluctuation records of palaeotimes are reconstructed by the aid of historical records through comparison of maps and photographs of different dates. The documented fluctuation records in Alps date back to 500 years and in the Himalayas they date back to only 150 years.[42, 43] The time duration of fluctuations has been variable.[44, 45, 46, 47, 48, 49, 50, 51] Recently Zumbuhl has reconstructed the fluctuation records and a glacial history of Grindelwald glaciers in Swiss Alps for 800 years from the twelfth century to the present. The reconstruction is based on the historical records in the form of realistic visual representation such as paintings, sculptures, photographs, and illustrated documents.[52, 53] This method of study is the first of its kind in the field of glacial geomorphology.

The fluctuation records of the snouts of the investigated glaciers are available for the past 100 years (A.D. 1857 to 1961).[54, 55, 56] The records reveal that the Kolahoi glacier snout has retreated up by 800 metres during A.D. 1857 to 1909 and another 800 metres during A.D. 1912 to 1961 and by 292 metres from the snout position of 1961 to that of 1984 (Figs. 30 and 31). The fluctuations have been documented on the basis of realistic visual representation like photographs, illustrated documents, and field observation[57, 58] (Figs. 31, 32 and 34). The Shishram glacier has retreated by 501 metres during A.D. 1923 to 1984 (Fig. 33).[59] This is similar to overall retreat of glaciers in the Himalaya. Their average retreat per year varies from 23.5 metres at Pindari to 4.6 metres in Potting, but there is a general decrease in the extent of retreat from the west to the east in the Himalaya ranging from 11.46 metres to 7.36 metres, respectively (Table 15).

Table 12 : *Steps Involved in Calculating Net Ablation Rate from Discharge of the East Liddar River at Chandanwari*

	365 days 1st Jan. to 31st Dec.	Winter period				Summer period					
		182 days 1st Oct. to 31st March	92 days 1st Oct. to 31st Dec.	90 days 1st Jan. to 31st March	30 days 15th Dec. to 14th Jan.	183 days 1st April to 30th Sept.	91 days 1st April to 30th June	92 days 1st July to 30th Sept.	30 days 15th May to 14th June	30 days 15th June to 14th July	30 days 15th July to 14th Aug.
	1	2	3	4	5	6	7	8	9	10	11
1. RT (mm)	1370 mm	662	199	463	204.5	708	478	230	164.8	162.3	98.5
2 R' (mm)	3.7	3.6	2.16	5.15	6.8	3.86	5.25	2.5	5.5	5.4	3.3
3 R1 = RTX25.6	0.35×10^{8}	0.169×10^{8}	0.051×10^{8}	0.119×10^{8}	0.052×10^{8}	0.181×10^{8}	0.12×10^{8}	0.059×10^{8}	0.042×10^{8}	$0.04.10^{8}$	0.025×10^{8}
R2 = RTX84.6	1.15×10^{8}	0.558×10^{8}	0.168×10^{8}	0.392×10^{8}	0.172×10^{8}	0.6×10^{8}	0.40×10^{8}	0.195×10^{8}	0.14×10^{8}	0.137×10^{8}	0.083×10^{8}
R3 = RTX130.8	1.79×10^{8}	0.863×10^{8}	0.26×10^{8}	0.60×10^{8}	0.267×10^{8}	0.926×10^{8}	0.62×10^{8}	0.30×10^{8}	0.216×10^{8}	0.21×10^{8}	0.128×10^{8}
4 Q (m^3/sec)	12.0	5.57	5.08	6.06	4.94	18.43	20.05	16.82	22.82	23.6	20.96
5 QT (m^3)	3.78×10^{8}	0.88×10^{8}	0.40×10^{8}	0.47×10^{8}	0.127×10^{8}	2.91×10^{8}	1.57×10^{8}	1.34×10^{8}	0.59×10^{8}	0.61×10^{8}	0.54×10^{8}

(Contd.)

Table 12—*Contd.*

	1	2	3	4	5	6	7	8	9	10	11
6 Q1 = QT-R1	3.43×10^{8}	0.771×10^{8}	0.349×10^{8}	0.351×10^{8}	0.075×10^{8}	2.729×10^{8}	1.45×10^{8}	1.281×10^{8}	0.548×10^{8}	0.57×10^{8}	0.515×10^{8}
Q2 = QT-R2	2.63×10^{8}	0.322×10^{8}	0.232×10^{8}	0.078×10^{8}		2.31×10^{8}	1.17×10^{8}	1.145×10^{8}	0.45×10^{8}	0.473×10^{8}	0.457×10^{8}
Q3 = QT-R3	1.99×10^{8}	0.017×10^{8}	0.14×10^{8}	—		1.984×10^{8}	0.95×10^{8}	1.04×10^{8}	0.374×10^{8}	0.40×10^{8}	0.412×10^{8}
7 A1 = Q1/25.6	13.4	2.8	1.37	1.36	0.3	10.6	5.7	5.0	2.15	2.2	2.0
A2 = Q2/25.6	10.27	1.25	0.9	0.3	—	9.02	4.6	4.5	1.8	1.8	1.8
A3 = Q3/25.6	7.8	0.07	0.6	—	—	7.75	3.7	4.1	1.5	1.6	1.6
8 A1/day cm	4.0	1.5	1.3	1.5	1cm	5.7	6.2	5.4	7.0	7.3	6.7
A2/day cm	3.0	0.7	0.9	0.3	less than 0.1	5.0	5.0	4.8	6.0	6.0	6.0
A3/day cm	2.0	—	0.6	—		4.2	4.0	4.5	5.0	5.3	5.3
	3.0	0.8	0.9	0.7	0.1	5.0	5.1	4.9	6.0	6.2	6.0

Source : Calculated from the discharge and rainfall data of Chandanwari (East Liddar River).

RT - Total Rainfall
R - Average Rainfall
R1 - Rainfall volume of ablation zone
R2 - Rainfall volume of Rain zone
R3 - Rainfall volume of entire Zone

Q - Average discharge of river basin
QT - The discharge volume of the river basin
Q1 - Volume of melt water flowing through ablation zone
Q2 - Volume of melt water flowing through rain zone
Q3 - Volume of melt water flowing through entire zone

A1 - Water equivalent of ablatien zone
A2 - Water equivalent of rain zone
A3 - Water equivalent of entire zone
A1/day - Ablation rate per day in ablation zone
A2/day - Ablation rate per day in rain zone
A3/day - Ablation rate per day in entire zone

Table 13 : *Steps Involved in Calculating Net Ablation Rate from the Discharge of the West Liddar River at Aru.*

	Winter period										*Summer period*
	365 days 1st Jan. to 31st Dec.	*182 days 1st Oct. to 31st March*	*92 days 1st Oct. to 31st Dec.*	*90 days 1st Jan. to 31st March*	*30 days 15th Dec. to 14th Jan.*	*183 days 1st April to 30th Sept.*	*91 days 1st April to 30th June*	*92 days 1st July to 30th Sept.*	*30 days 15th may to 14th June*	*30 days 15th June to 14th July*	*30 days 15th July to 14th Aug.*
	1	2	3	4	5	6	7	8	9	10	11
1. RT mm	1370	662	169	463	204.5	708	478	230	164.8	162.3	98.5
2. R	3.7	3.6	2.16	5.15	6.18	3.86	5.25	2.5	5.5	5.4	3.3
3. $R1 = RT \times 22.4$	0.31×10^8	0.14×10^8	0.45×10^8	0.1×10^8	0.045×10^8	0.16×10^8	0.1×10^8	0.05×10^8	0.37×10^8	0.036×10^8	0.02×10^8
$R2 = RT \times 185.6$	2.54×10^8	1.21×10^8	0.37×10^8	0.85×10^8	0.37×10^8	1.31×10^8	0.88×10^8	0.42×10^8	0.3×10^8	0.3×10^8	0.18×10^8
$R3 = RT \times 260.5$	3.56×10^8	1.72×10^8	0.51×10^8	1.2×10^8	0.53×10^8	1.84×10^8	1.24×10^8	0.6×10^8	0.43×10^8	0.42×10^8	0.25×10^8
4. Q (m^3/s)	16.6	7.76	9.2	6.3	7.4	25.43	30	20.86	34.8	28.74	23.65
5. QT (m^3)	5.23×10^8	1.22×10^8	0.73×10^8	0.49×10^8	0.19×10^8	4.0×10^8	2.35×10^8	1.65×10^8	0.8×10^8	0.74×10^8	0.61×10^8

(Contd.)

Table 13 :—*Contd.*

	1	2	3	4	5	6	7	8	9	10	11	
6. Q1 = QT-R1	4.92×10^{8}	1.06×10^{8}	0.685×10^{8}	0.49×10^{8}	0.145×10^{8}	3.84×10^{8}	2.25×10^{8}	1.60×10^{8}	0.863×10^{8}	0.704×10^{8}	0.59×10^{8}	
Q2 = QT-R2	2.69×10^{8}	0.01×10^{8}	0.40×10^{8}	—	—	2.69×10^{8}	1.47×10^{8}	1.23×10^{8}	0.6×10^{8}	0.44×10^{8}	0.43×10^{8}	
Q3 = QT-R3	1.67×10^{8}	0.22×10^{8}	2.16×10^{8}	1.15×10^{8}	1.05×10^{8}	0.47×10^{8}	0.32×10^{8}	0.36×10^{8}				
7. A1 = Q1/22.4	21.96	4.73	3.0	1.74	0.64	17.24		10.10	7.14	3.85	3.14	2.63
A2 = Q2/22.4	12.0	0.04	1.8	—	—	12.0	6.56	5.5	2.68	1.96	1.92	
A3 = Q3/22.4	7.5	—	nearly 1.0		—	—	9.6	5.14	4.69	2.1	1.43	1.60
8. A1/days cm	6.0	2.5	3.2	2.0	2.1	9.4	11.0	7.8	12.8	10.5	8.8	
A2/days cm	3.3	—	2.0	—	—	6.5	7.2	6.0	9.0	6.5	6.4	
A3/days cm	3.0	—	1.1	—	—	5.2	5.6	5.0	7.0	4.8	5.3	
MEAN	3.8	2.5	2.1	2.0	2.1	7.0	7.9	6.3	9.6	7.3	6.8	

Source : Calculated from the discharge of the West Liddar River at Aru and from rainfall data of Chandanwari.

Table 14 : *Steps Involved in Calculating Net Ablation Rate from Discharge of the East Liddar River at Shishnag.*

	Winter period					*Summer period*					
	365 days 1st Jan. to 31st Dec.	*182 days 1st Oct. to 31st March*	*92 days 1st Oct. to 31st Dec.*	*90 days 1st Jan. to 31st March*	*30days 15th Dec. to 14th Jan.*	*183 days 1st April to 30th Sept.*	*91 days 1st April to 30th June*	*92 days 1st July to 30th Sept.*	*30 days 15th May to 14th June*	*30 days 15th June to 14th July*	*30 days 15th July to 14th Aug.*
	1	2	3	4	5	6	7	8	9	10	11
1. RT	1370 mm	662 mm	199 mm	463	204.5	708	478	230	164.8	162.3	98.5
2. R	3.7 mm	3.6 mm	2.16	5.15	6.8	3.86	5.25	2.5	5.5	5.4	3.3
3. R1 = RTX13	0.176×10^{8}	0.085×10^{8}	0.025×10^{8}	0.06×10^{8}	0.02×10^{8}	0.09×10^{8}	0.06×10^{8}	0.03×10^{8}	0.02×10^{8}	0.02×10^{8}	0.012×10^{8}
R2 = RTX25	0.34×10^{8}	0.15×10^{8}	0.047×10^{8}	0.12×10^{8}	0.05×10^{8}	0.57×10^{8}	0.12×10^{8}	0.067×10^{8}	0.04×10^{8}	0.04×10^{8}	0.025×10^{8}
R3 = RTX55.5	0.75×10^{8}	0.36×10^{8}	0.1×10^{8}	0.25×10^{8}	0.097×10^{8}	0.38×10^{8}	0.27×10^{8}	0.13×10^{8}	0.88×10^{8}	0.09×10^{8}	0.054×10^{8}
4. Q (m^3/s)	8.5	4.56	4.0	5.1	3.8	12.5	12.8	12.3	15.4	15.8	14.2
5. QT (m^3)	2.68×10^{8}	0.72×10^{8}	0.32×10^{8}	0.097×10^{8}	0.098×10^{8}	1.98×10^{8}	1.0×10^{8}	0.98×10^{8}	0.4×10^{8}	0.41×10^{8}	0.37×10^{8}

(Contd.)

Table 14 : *Contd*

	1	2	3	4	5	6	7	8	9	10	11
6. Q1 = QT-R1	2.504×10^{8}	0.635×10^{8}	0.295×10^{8}	0.337×10^{8}	0.078×10^{8}	1.89×10^{8}	0.94×10^{8}	0.95×10^{8}	0.38×10^{8}	0.39×10^{8}	0.36×10^{8}
Q2 = QT-R2	2.34×10^{8}	0.57×10^{8}	0.273×10^{8}	0.277×10^{8}	0.048×10^{8}	1.81×10^{8}	0.88×10^{8}	0.92×10^{8}	0.36×10^{8}	0.37×10^{8}	0.34×10^{8}
Q3 = QT-R3	1.93×10^{8}	0.36×10^{8}	0.22×10^{8}	0.147×10^{8}	—	1.60×10^{8}	0.73×10^{8}	0.85×10^{8}	0.31×10^{8}	0.32×10^{8}	0.32×10^{8}
7. A1 = Q1/13	19.23	4.85	2.23	2.6	—	14.5	7.23	7.3	2.92	3.0	2.77
A2 = Q2/13	18.0	4.4	2.1	2.1	0.6	13.9	6.76	7.08	2.77	2.84	2.6
A3 = Q3/13	14.84	2.8	1.7	1.15	0.38	12.3	5.6	6.5	2.38	2.46	2.46
8. A1/day cm	5.2	2.6	2.4	2.9	2.0	7.9	7.9	7.9	9.7	10.0	9.2
A2/day cm	4.9	2.4	2.2	2.3	1.2	7.5	7.4	7.6	9.2	9.4	8.6
A3/day cm	4.0	1.5	1.8	1.3	—	6.7	6.2	7.1	7.9	8.2	8.2
MEAN	4.7	3.2	2.1	2.2	1.6	7.4	7.2	7.5	8.9	9.2	8.9

Source : Calculated from Discharge of East Liddar river at Shishnag and from the Rainfall Data of Chandanwari.

Fig. 30: *The Position of Snout of the Kolahoi Glacier in 1909 is near the Kolahoi Ganj Area. It is nearly 1* ***Kilometre*** *Northwest of the 1985 Snout Position. (Photograph taken by Neve in 1909 and is Reproduced from the Alpine Journal).*

Table 15: *Average Retreat in the Himalayan Glaciers*

Glacier	*Period*	*Years*	*Retreat (metres)*	*Average Retreat (metres/year)*
Pindari	1845 - 1966	121	2840	23.5
Gangotri	1935 - 1976	41	600	14.6
Kolahoi	1857 - 1984	127	1792	14.1
Millam	1849 - 1957	108	1420	12.5
Machoi	1906 - 1957	51	457	9.0
Shishram	1923 - 1984	61	501	8.4
Zemu	1909 - 1965	56	440	7.8
Potting	1906 - 1957	57	262	4.6
Liddar	1984 - 1985	1	3	3.0
Nahanar	1976 - 1977	1	2	2.0

Source: C.P. Vohra (1981), "Himalayan Glaciers", *Himalaya Aspects of Change*, Lall, J.S. and Moddie, A.D. (eds.) New Delhi; Oxford University Press, p. 141.

Fig. 31: *The Photograph of Kolahoi Glacier Snout taken by Author in 1985. Marks the present position of the glacier snout (A). The principal feeder of the glacier derives from the west side of the peak marked by arrow. Position of the glacier snout in 1961 (B) as recorded by Odell is near the lower reaches of the mountain. The location of the lateral moraines (C) formed by the recession of 1961 ice mass marks the extent of retreat of snout since August 1961.*

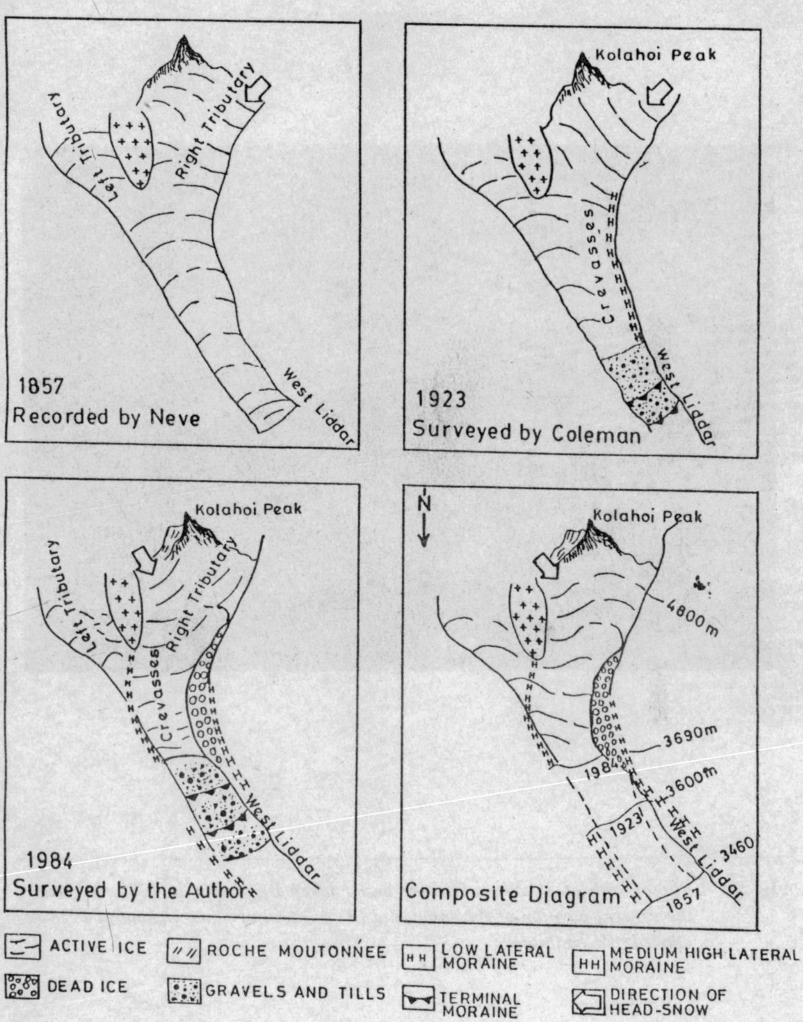

Fig. 32: *Fluctuations in the Snout Position of the Kolahoi Glacier (1857-1984)*

Fig. 33: *Photograph of Shishram Glacier Snout taken by Author in 1985. (A) marks the present position of the snout and (B) position of glacier snout in 1928 recorded by Grinlinton*

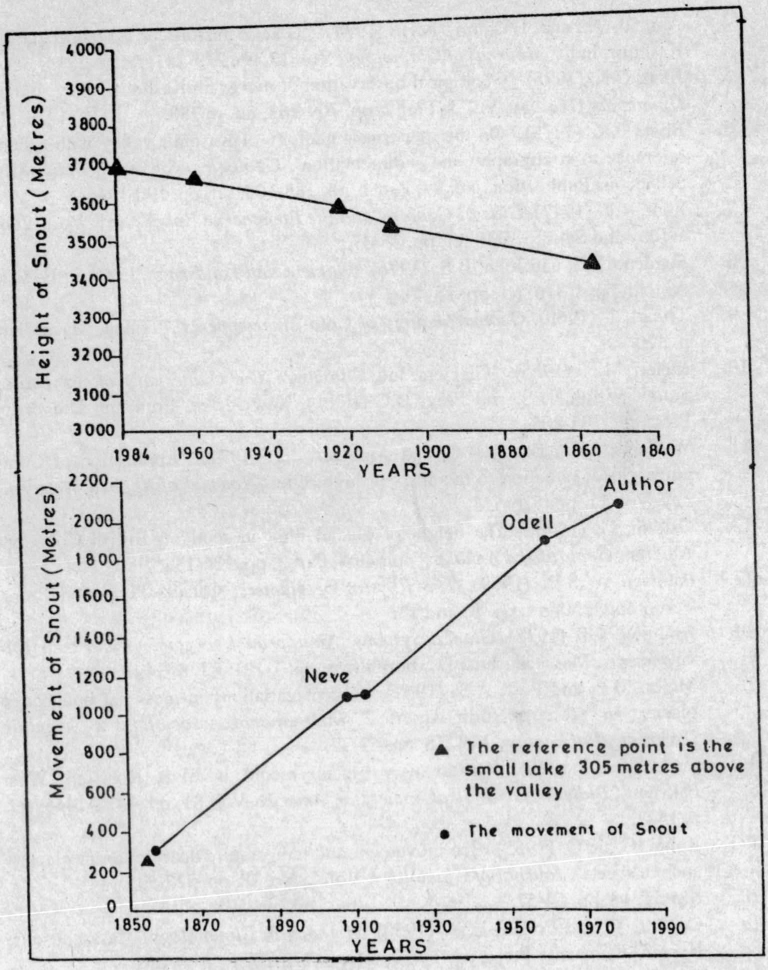

Fig. 34: *Retreat of Kolahoi Snout*

NOTES

1. Ahmad, N. and Mayeswki, P.A. (1981). 'Himalayan Glaciation: An Unresolved Issue', *The Geographer*, Vol. 28, No. 2, pp. 21 - 32, ref. p. 27.
2. Vivian, R. (1975). *Less glaciers des Alpes Occidentales*, Grenoble; Allier, pp. 1-513.
3. Grinlinton, J.L. (1928). 'Former glaciation of the East Liddar Valley, Kashmir', *Memoir, Geological Survey of India*, Vol. 49, Part 2, pp. 289-388.

4. Ahmad, N. and Hashimi, N.H. (1974). 'Glacial history of Kolahoi glacier, Kashmir, India', *Journal of Glaciology*, Vol. 13, pp. 279-283, ref. p. 281.
5. Bhat, D.K. (1978). 'Geological observation of *margs* on Kashmir valley, India', *Himalayan Geology*, Vol. 8, Part 2, pp. 769-783, ref. p. 780.
6. Bhat, D.K. (1975). 'On the quaternary geology of Kashmir valley with special reference to stratigraphy and sedimentation', *Geological Survey of India*, Miscellaneous Publication, No. 24, Part 1, pp. 188-203, ref. pp. 190-194.
7. Flint, R.F. (1947). *Glacial Geology and the Pleistocene Epoch*, New York, John Willey and Sons, p. 560, ref. pp. 35-37.
8. Sugden, D.E., and John, B.S. (1976). *Glaciers and Landscape*, London; Edward Arnold, pp. 1-376, ref. pp. 15-40.
9. Tricart, J. (1969). *Geomorphology of Cold Environments*, London, Macmillan, p. 320.
10. Meier, M.F. (1965). 'Glaciers and Climates', *The Quaternary of the United States*, Wrigh, H.E. and Frey, D.G. (Eds.), New Jersey, Princeton University Press, pp. 795-805.
11. Williams, L.D., Barry, R.G. and Andrews, J.T. (1972). 'Application of Computed global radiation to the areas of high relief', *Journal of Applied Meteorology*, Vol. 11, pp. 526-533.
12. Ostrem, G. (1966). 'The height of glacial limit in southern British Columbia Alberta', *Geografiska Annaler*, Vol. 48A, Part 3, pp. 126-138.
13. Paterson, W.S.B. (1969). *The Physics of glaciers*, Canada Pergamon Press, Toronto; p. 250, ref. pp. 12 and 13.
14. Andrews, J.T. (1975). *Glacial Systems: An approach to glaciers and their Environments*, Massachusetts; Duxbury Press, pp. 1-191, ref. p. 34.
15. Meier, M.F. and Post, A.S. (1962). 'Recent variations in mass net budgets of glaciers in Western North America', *International Association of Scientific Hydrology Publication*, Vol. 58, pp. 63-77.
16. Porter, S.C. (1970). 'Quater nary glacial record in Swat Kohistan, West Pakistan', *Bulletin, Geological Society of America* Vol. 81, pp. 1421-1446, ref. p. 1443.
17. Robin, G. de Q. (1955). 'Ice movement and temperature distribution in glaciers and ice sheets', *Journal of Glaciology*, Vol. 2, No. 18, pp. 523-532.
18. See ref. 13, pp. 54-57.
19. Conard, V. and Pollak, L.W. (1950). *Methods in Climatology*, Massachusetts; Harvard University Press, pp. 1-459, ref. pp. 272-282.
20. See ref. 7, pp. 72-74
21. Hoinkes, H.C. and Rudolph, R. (1962). 'Mass Balance studies on Hinterisferner, Otzal Alps, 1951—61', *Journal of Glaciology*, Vol. 4, No. 33, pp. 260-280.
22. Grosval'd, M.G. and Kotlyakov, V.M. (1969). 'Present-day glaciers in the U.S.S.R. and some data on their mass balance', *Journal of Glaciology*, Vol. 8, pp. 9-22.
23. Meier, M.F. and Taugborn, W.V. (1965). 'Net budget and flow of South Cascade Glacier, Washington', *Journal of Glaciology*, Vol. 5, pp. 547-566.
24. Ostrem, G. (1966). 'Mass balance studies on glaciers in Western Canada 1965', *Geographical Bulletin*, Vol. 8, pp. 81-107.
25. Schytt, V. (1962). 'Mass balance Studies in Kebnekajse', *Journal of Glaciology*, Vol. 4, pp. 281-288.

26. Tewari, A.P. (1973). 'Recent changes in the position of the snout of Pindari glacier (Kumaon Himalayas) Almora district, U.P., India', *The Role of Snow and Ice in Hydrology Proceedings of the Banff Symposia*, Vol. 2, Unesco-WMO-IAHS, pp. 1144-1149.
27. Mayeswki, P.A. and Ahmad, N. (1981). 'Reconnaissance glacio-chemical studies in the Indian Himalayas', *Proceedings of Eastern Snow Conference*, 38th Annual meeting, New York, pp. 45-48.
28. Raina, V.K. and Kaul, M.K. and Singh, S. (1977). 'Mass balance studies of Gara glacier', *Journal of Glaciology*, Vol. 18, No. 80, pp. 415-423.
29. Bhandari, N., and Nijampurkar, V.N. (1981). 'Glaciers: Our neglected water banks', *Science Today*, No. 12, pp. 60-62.
30. Nye, J.F. (1965). 'A numerical method of inferring the budget history of glacier from its advance and retreat', *Journal of Glaciology*, Vol. 5, pp. 589-607.
31. Schytt, V. (1967). 'A Study of Ablation gradient', *Geografiska Annaler*, Vol. 49A, Number 2-4, pp. 327-332.
32. See ref. 13, pp. 56-57.
33. Ostrem, G. and Stanely, A.D. (1966). *Glacier mass balance measurements, A manual for fieldwork*, Canada: Department of Mines and technical surveys, Ottawa.
34. Mokievsky, Z.O. (1974). Analysis of mass balance values and their accuracy for Sentinal glacier, British Columbia Canada', *Inland water directorate resource branch*, scientific series No. 31.
35. Ahlman, H.W. (1948). 'Glaciological research on the North Atlantic Coast', *Royal Geographical Society Research Series* 1, pp. 1-83.
36. King, C.A.M. (1978). Techniques in Geomorphology, *London: Edward Arnold*, pp. 1-341, ref. p. 95.
37. See ref. 25, p. 285.
38. Meier, M.F. (1962). 'Proposed definitions of glacier mass budget term', *Journal of Glaciology*, Vol. 4, pp. 252-261.
39. Paterson, W.S.B. and Savage, J.C. (1963). 'Geometry and movement of Athabasca glacier', *Journal of Geophysical Research*, Vol. 68, No. 15, pp. 4513-4520.
40. See ref. 13, pp. 38-39.
41. Mayeswki, P.A. and Pregent, G. (1978). *Results of the 1978 Athabasca glacier expedition*, New Hampshire, Department of Earth Sciences, University of New Hampshire, p. 56, ref. pp. 23-26.
42. Mayeswki, P.A. and Jaschke, P.A. (1979). 'Himalayan and Trans-Himalayan glacier fluctuations since A.D. 1812', *Arctic and Alpine Research*, Vol. 11, No. 3, pp. 267-287, ref. p. 267.
43. Tewari, A.P. (1971). 'A short report on glacier studies in Himalayan mountains by the Geological Survey of India', *Stud. Geomoorphol. Carpatho-Balcancia*, Vol. 5, pp. 173-181.
44. *Geographical Journal* (1925). 'The Movement of the Indian Glaciers', Vol. 66, Part 4, pp. 331-334.
45. Hayden, H.H. (1907). 'Notes on certain glaciers in Northwest Kashmir', *Record, Geological Survey of India*, 35, pp. 127-137.
46. Raina, V.K. (1963). 'A note on some glaciological observations in Garhwal Himalayas', *Indian Minerals*, Vol. 17, Part 2, pp. 159-163.

47. Raina V.K., Bhattacharya, V. and Pattnaik, S. (1973). 'Zemu Glacier', *Record, Geological Survey of India*, Vol. 105, No. 2, pp. 95-106.
48. Tewari, A.P. and Jangpangi, B.S. (1962). 'The retreat of Snout of Pindari glacier', in *Variations of the regime of existing glaciers symposium, commission of snow and ice*, U.S.A., Association of Scientific Hydrology Publication, Vol. 58, pp. 245-246.
49. Strachey, R. (1848). 'Notes on the motion of glaciers of Pindur in Kumaon', *Journal of the Asiatic Society of Bengal*, Vol. 17 pp. 203-205.
50. Odell, N.E. (1963). 'The Kolahoi Northern Glacier, Kashmir', *Journal of Glaciology*, Vol. 4, pp. 633-635.
51. Benney, T.G. (1908). 'Fluctuations in Himalayan glaciers', *Nature*, Vol. 77, pp. 201-202.
52. Zumbuhl, H.J. (1980). *Die Schwankungen der Grindelwaldgletscher: in den historischen Bild-und Schrifiquellen des 12. bis 19. Jahrhunderts*, Basel, Birkhauser Verlag, pp. 1-187.
53. Messerli, B., Messerli, P., Pfister, C. and Zumbuhl, H.J. (1978). 'Fluctuations of climate and glaciers in the Bernese Oberland, Switzerland and their geoecological significance, 1600 to 1975', *Arctic and Alpine Research*, Vol. 10, No. 2, pp. 247-260.
54. Mayewski, P.A., Pregent, G.P. Jeschke, P.A. and Ahemad, N. (1980). 'Himalayan and Trans-Himalayan glacier fluctuation and the South Asian monsoon record', *Arctic and Alpine Research*. Vol. 12, No. 2, pp. 171-182.
55. See ref. 42, p. 273.
56. Neve, E.F. (1910). 'Mt. Kolahoi and its northern glacier', *Alpine Journal*, Vol. 25, pp. 39-42.
57. See ref. 50, p. 634.
58. See ref. 3, pp. 41-42.
59. See ref. 3, pp. 294-295.

4

Glacial Erosion: Processes and Geomorphological Features

GLACIAL GEOMORPHOLOGY

Glacial geomorphology is the study of landforms produced by glacial and fluvioglacial processes in areas of present glaciers as well as in areas covered by glaciers in the Pleistocene Epoch. It is concerned with the classification, measurements, and description of the landforms. The glacial landforms are intimately related to the total mass budget of the glacier system. It determines the degree of glacial acitivity and therefore controls the glacial process. The mass budget indicates the relationship between mass gain in the accumulation zone in higher elevation and mass loss due to melting in the ablation zone of the glacier, as discussed in the previous chapter. The accumulation and ablation zone in a glacier are separated by geomorphic threshold. The transfer of threshold from valleyhead indicates the basal slippage of the glacier.[1] It results in sending action of glacial base due to mass transfer of energy by advance and retreat of the glacier. The advance and retreat in their own turn generate the glacial processes which create erosional and depositional features in glacial geomorphic landscape.

Much of the spectacular character of the present topography of the northern part of the Liddar valley has resulted from the glaciation in the remote past. A study of the glacial processes operating currently can throw light on the landforms that were created by similar processes in the past. Therefore, it will be helpful in reconstructing the genesis of the landform features.

Himalayan glacial landscape has received very little attention from geomorphologists despite its extensive record of Pleistocene glaciation. To a large extent this is the result of formidable difficulties posed by the remote mountainous terrain. Hardly any work on the geomorphology of the study region is available except the reconnaissance survey published by Grinlinton and glacial history documented by Ahmad and

Hashimi.[2, 3] It does not throw any light on the genesis of the landform features. The present study attempts to produce an explanatory descriptive analysis of the landform features.

PREGLACIAL LANDFORM

The Liddar valley glaciation owes its existence to the preglacial topography carved by rejuvenated river system. Preglacial drainage pattern appears to have been largely conformable with structure and normal denudation. The carved drainage channels of various sizes were the main routes of the glacial advance and retreat. During the advance of the glaciers the rock walls of the East and the West Liddar basins became partly submerged. There were three advances followed by retreat. The third advance was an extensive one. It resulted in high mass transfer from valleyhead, at an altitude of 5000 metres above sea level, to a down glacier position at an altitude of 2000 metres above sea level. It was followed by negative feedback of glaciers, which resulted in the retreat in three phases.

LIDDAR VALLEY : A POLYGENETIC LANDFORM

Though the valley reveals clear evidence of glacial activity, the exact importance of ice erosion for development of specific forms may be difficult to assess. At certain places near the valleyheads of the East and the West Liddar the plantation levels have been glaciated and modified by ice action. A major part of the U-shaped trough carved in the East and the West Liddar valley above an altitude of 3600 metres consists of steep clear-cut water-face slope which has created a regional step in the landform. The upper lip of the step is buried under the present ice. Grinlinton has observed a clear-cut water-run slope near Shishnag lake.[4] It is located at a height of 260 metres from the lake. It is in agreement with the observation made during the present investigation. The present valley landscape, therefore, is polygenetic, composed of the elements produced by glacial, fluvio-glacial and fluvial agents. Large parts of the valley below the shoulders of Kolahoi, Basmai, Kunyirhayan, and Katsal mountains have been formed by preglacial erosion. The preglacial drainage evolved through capture and countercapture in valleyhead at Nafronnar and Chalwasnar. Between the altitudes of 2000 and 3500 metres the deeper part of the valley reveals an alternate sequence of overdeepening the reglaciation. Below the altitude of 2000 metres the landscape is primarily fluvial in nature. Glacial action has modified the long profile of the northern part of the Liddar valley but not the main drainage lines. The knolls near Chandanwari, Aru, Mundlun, Pharsulun and Pahalgam in the valley bottom, have not been changed.

The landform produced by glacial erosion, transportation or deposition varies at different sites of the valley from Ganshibal to Kolahoi Ganj and Shishnag lake due to variation in altitude variation in lithology of glacier bed, thickness of ice and its fluctuation in mass balance. These parameters have promoted the morphological activity of landform formation. The character of glacier erosion depends upon the thickness of ice, speed of ice movement and time available for glacial process to act on landform. The glacial transport is mainly governed by slope of the landform and velocity of ice mass carrying load. The material transported by it consists of supraglacial material that falls into ice surface from nearby valley sides, cirques and material scoured from the bed. The glacial deposition depends upon ice temperature, velocity of glacier and configuration of bed.

GLACIAL EROSION

Most of the glacial sculpturing manifest in the Liddar mountains was effected during Ganshibal stade. During this stade glaciers ironed out the sharp relief which was inherited from preglacial period. Erosion was guided by structure of the landform and in a general way this has continued to the present. Most of the ranges and major components thereof correspond closely to structural trend and their crests coincide reasonably well with anticline. As a result of long continued glacial erosion, mountains are underlain by synclines and valleys by anticlines. The glacial landforms suggests that there is a steady decline in the amount and change in the form of glacial erosion from north to south. The intensity of the excessive glacial erosion is confined to the middle section and reduced erosion to the upper and the lower sections of the East and West Liddar valley (Fig. 41). In the middle section of the valley a large number of rock basins, whalebacks and roche moutonnees are found and striation on them suggest that the excessive erosion is confined to those areas that lay parallel to ice movement. Further, the long profile of the East and the West Liddar valley also reveals a stepped form in the headward part from Pissu Ghati to Shishnag lake and Kalwan to Kolahoi Ganj, respectively. Further down the steps are rare and there are only rock basins. The steps have been observed at Pissu Ghati, Zaijpal fall and at Kalwan and Liddarwat fall in the East and the West Liddar valley respectively. The steps coincide with the poorly jointed parts of sandstone and basalt, which being massive are resistant to erosion. The rock basins are related to moderately well jointed rocks and are produced by overdeepening of the valleys as revealed in Pharsulan, Chandanwari, Zaijpal, Shishnag and Aru, Satlanjan and Kolahoi in the East and the West Liddar, respectively. The deepest rock basins are located at Zaijpal (650 metres deep) and Liddarwat (725 metres deep) near the middle section of the Liddar valley.

EROSIONAL PROCESSES

Abrasion and plucking are the two main processes responsible for the creation of the erosional landforms of the study region. They operated when the glaciers moved in bedrock channels of the East and the West Liddar valley in response to gravity and basal sliding (Fig. 35). The basal sliding of the glaciers accounted for nearly 90 per cent of the total movement because of steep slope in the catchment area. Through their movements the glaciers applied pressure on the contact areas of valley floor and valleysides. Their movement was greatly affected by frictional retardation of undulated bedrock. There was frictional drag between the sole of the glacier and bedrock. Due to this drag potential energy was created by glacier. This energy is similar to shear stress and is expressed as:

$$r = \frac{egh}{A} \times \sin \theta$$

Where

r = shear stress
e = density of ice
g = gravitation acceleration
h = thickness of ice
A = area of contact
$\sin \theta$ = slope of ice

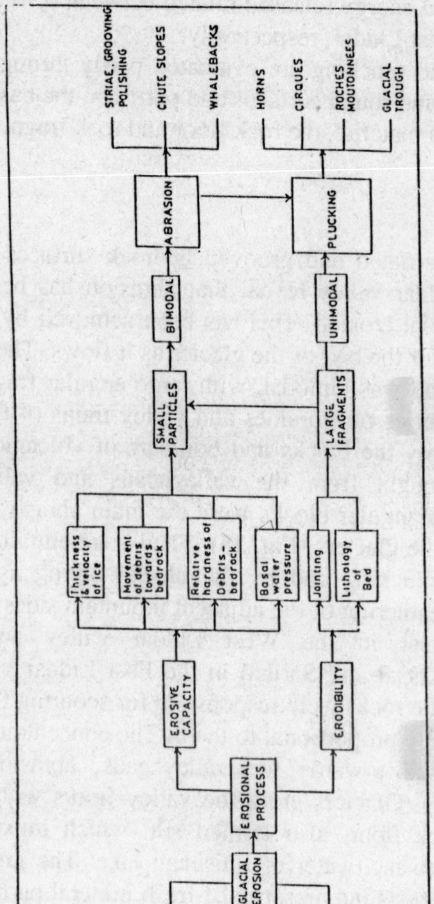

Fig. 35: *Model of Liddar Glacier Erosion and Erosional Landform*

The potential energy released ranged between 0.75 and 0.78 bars in the West and East Liddar, respectively.

Abrasion and plucking are evaluated partly through an interpretation of smooth and rough surfaces and partly on the basis of quantity of glacially eroded material, the rock floor and rock fragments.

Abrasion

The polished, striated and grooved bedrock surfaces of the northern part of the Liddar valley reveal that abrasion has been an important process of glacial erosion. This has been achieved by the dragging of the particles over the bed of the glacier as it flows. The particles varied in size. The load was bimodal, with large angular fragments and finer debris. The debries of moraines and valley trains of the Liddar valley contain primarily the blocks and boulders of volcanic and calcareous sandstones brought from the valleyheads and valleywalls by the glaciers. These angular blocks were the main abrasive tools occurring at the sole of the glaciers (Fig. 35). They were entrained in the glacier sole through plucking and by the blocks falling as a result of the mechanical weathering of the adjacent mountain sides of Kolahoi, Basmai and Tarsar in the West Liddar valley and Kunyirhayan, Mahagunus, Katsal and Saribal in the East Liddar valley. The larger fragments of the rocks were responsible for scouring the land and abrasion was directly proportional to them. The concentration of large fragments increased towards the valleyheads, abrasion too increased proportionately. Glaciers grind the valley floors with large fragments producing rock flour, also termed silt, which imparts characteristic green colour to meltwater of Shishnag lake. The green colour in the meltwater suggests the presence of fresh mineral particles smaller than 100 micron diameter.[5] Its estimation is useful for finding out abrasion rate. In the study region the silt data are available for Shishnag lake only. In Shishnag the meltwater streams of Shishram glacier directly enters the lake. The gauge station is located at the mouth of the East Liddar river. The silt data closely correlate with the discharge of Shishnag lake. The maximum silt is formed in meltwater streams during the summer season and the minimum during the winter season (Fig. 36). During the summer season the silt concentration in one litre of water ranges between 0.029 and 0.035 cubic centimetres, while during the winter season its concentration decreases to 0.005 cubic centimetres (Table 16). The average silt load per litre was 0.021 cubic centimetres.

It amounted to 0.94 millimetres of erosion during the 1983-84 field seasons. From the rate one can suggest that abrasion alone is currently lowering the floor of Shishram glacier basin by about 1 metre every 1000 years (Table 16). This is analogous to overall lowering of glacier basin due to abrasion in the Liddar valley.

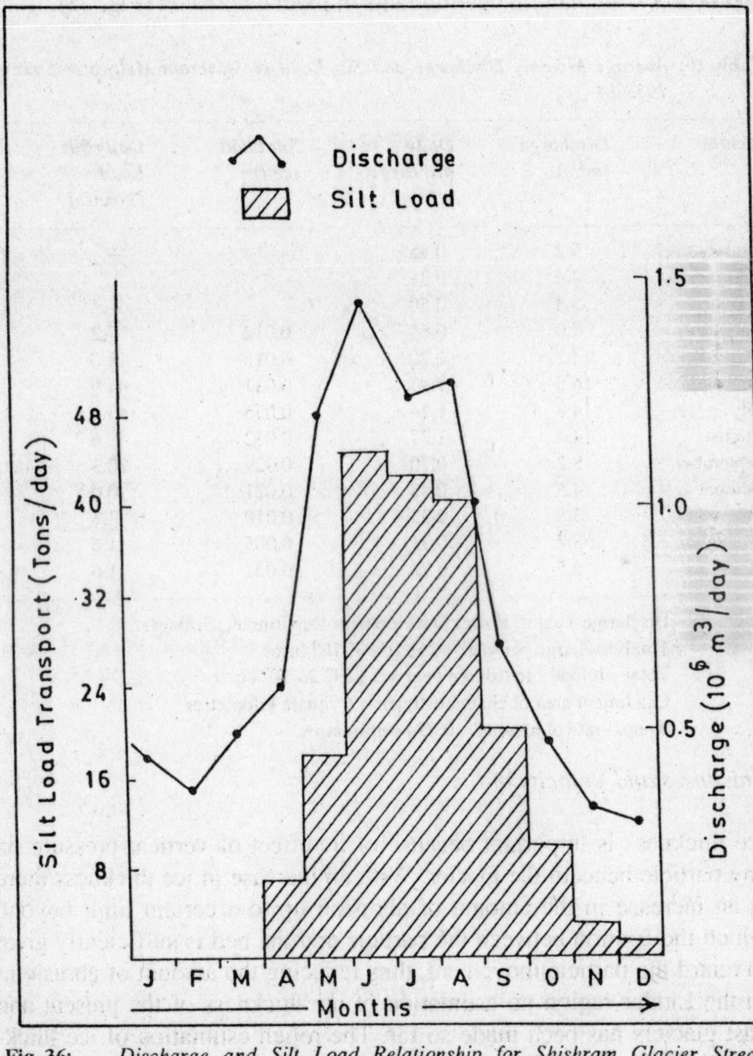

Fig. 36: *Discharge and Silt Load Relationship for Shishram Glacier Stream, 1983-84*

Determinants of Abrasion

Thickness and velocity of glacier, transport of debris towards the bedrock, basal water pressure, and relative hardness of rock particles and bedrock are natural determinants of glacial abrasion (Fig. 35).

Table 16: *Average Monthly Discharge and Silt Load in Shishram Meltwater Stream, 1983-84*

Months	*Discharge (m^3/s)*	*Daily discharge (10^6 m^3)*	*Silt Load (cc/l)*	*Daily Silt Load (Tonnes)*
January	5.2	0.45	—	—
February	4.4	0.38	—	—
March	5.8	0.50	—	—
April	7.0	0.60	0.012	7.2
May	14.2	1.20	0.015	18.0
June	16.8	1.45	0.031	44.9
July	14.3	1.24	0.035	43.4
August	14.4	1.27	0.032	40.6
September	8.2	0.70	0.029	20.3
October	4.5	0.48	0.021	10.0
November	3.8	0.33	0.010	3.3
December	3.7	0.31	0.005	1.6
Mean	8.5	0.73	0.021	1.6

Source: Discharge Tables, Power Development Department, Srinagar.
Total discharge per year = 268.056 X 10^9 litres
Total silt load carried in a year = 5.6292 X 10^9 cc.
Catchment area of channel/stream = 6 square kilometres
Annual rate of erosion = 0.094 centimetres.

Thickness and Velocity of Ice

Ice thickness is important because of its effect of vertical pressure on any particle beneath the glacier. With an increase in ice thickness there is an increase in the amount of abrasion up to a certain limit beyond which the friction between the particle and the bed is sufficiently great to retard the particle movement, thus reducing the amount of abrasion.[6] In the Liddar region no estimation of ice thickness of the present and past glaciers has been made so far. The rough estimation of ice thickness could be arrived at from the minor and major forms of glacial processes which had operated in the Liddar valley. In the headwalls of the East Liddar valley the striated topography has been observed nearly 250 metres above the Shishnag lake and the latter is about 15 metres deep. Such an anomalous juxtaposition in the landform of two features could have been created by huge mass of ice. Adjacent to the lake there is a Kame delta. It is nearly 200 metres above the valley floor. It is estimated that 250 metres thick ice had carved out the lake basin. The rock whalebacks and roche moutonnees, near Chandanwari, Kalwan and Pahalgam, are respectively 15 and 50 metres high from the valley beds. The three features and their dimensions provide us with the estimates of the magnitude of ice thickness responsible for abrasion. The ice thickness of 250 metres could have produced enormous pressure on the bedrock as well as on rock fragments in the glacier sole. Abrasion also depends on the speed of the glacier sliding, which controls the amount of debris passing over a given area of bedrock.[7] The average velocity of the present glaciers is nearly 3 metres per year and their normal pressure about 1.5 bars.[8] The velocity of ice during the active phase could have been much faster, perhaps at least twenty times faster than the present flow rate and pressure on the bedrock and basal fragments could have been nearly 30 to 50 bars, when the landforms were moulded.

Transport of Debris Towards the Bedrock Side

Abrasion by rock particles is apparently ineffective unless particles are continually moving towards the bed of the glacier and constantly renewing the process of abrasion. In the Liddar valley the melting of ice at the bottom of the glaciers due to warm temperature was the most effective way in which the fragments could have moved down towards the bed surface. It was further helped by an increased downward velocity of glaciers near the source of meltwater streams of the East and the West Liddar. The renewal of fragments at the sole of the glacier was effected by basal melting and inclined slope (Fig. 37). Abrasion requires the concentration of debris in the basal layer, as clean ice cannot abrade the solid rock.[9] The abrading potentialities of the Liddar glaciers could have increased tremendously after the third advance, because it has been retreating in phases, since then. The retreat has increased the frequency and volume of rock-fall. The rock-falls on to the glacier surface from the adjacent walls of cirque and valley glaciers. The process is generated by the pressure release on the fractures. The abundance of rock fragments and high thermal gradient

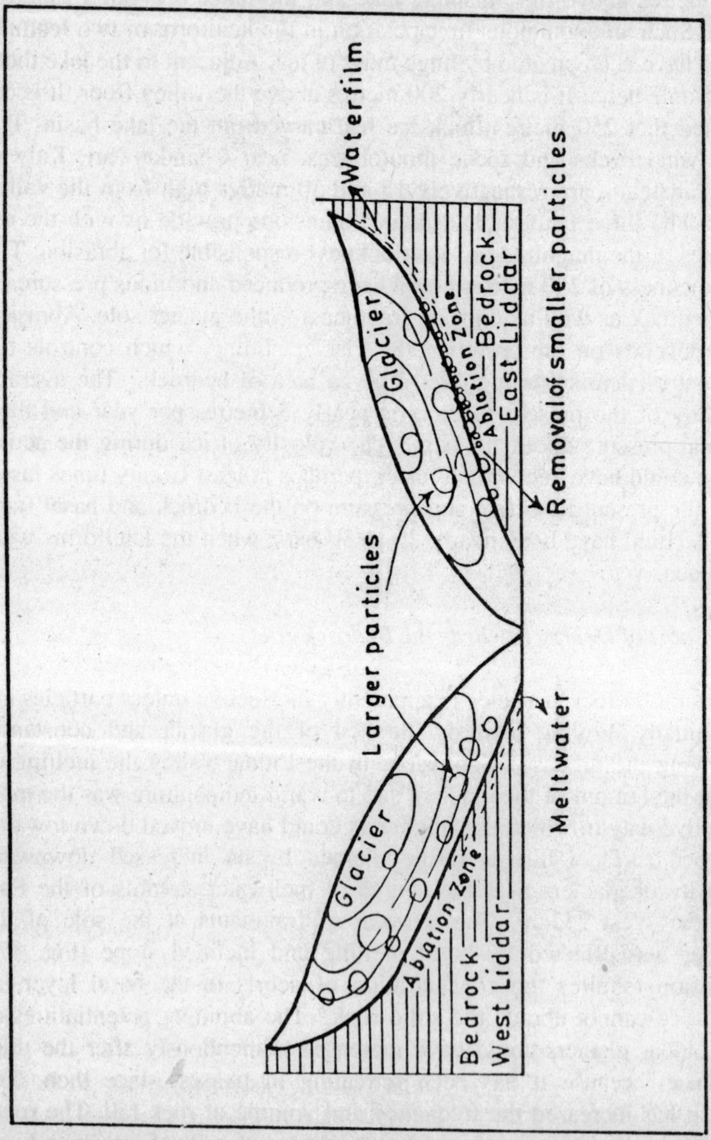

Fig. 37: *Removal of Smaller Particles by Meltwater*

at the base of ice has led to movement of fragments towards the glacier sole. The fragments, mostly being angular, have striated and polished the bedrock while the glacier is retreating.

Basal Water Pressure

The Liddar glacier is a warm-based glacier and has a thin layer of water at its base. The thin film of water acts as a lubricating agent and helps in basal sliding which enhances the abrasion power by transporting large fragments towards the glacier sole and removing the smaller particles through buoyancy effect.

Relative Hardness of Rock Particles and Bedrock

The relative difference in hardness of particles in the glacier sole and bedrock effect the character of erosion. The presence of basaltic and andesitic rocks near the valleyhead of the study region has promoted striation and polishing of the rocks, near the localities of Kolahoi Ganj and Shishnag lake. The sole of glacier contains hard resistant blocks of basalt and andesite which were dragged along the limestone rocks of Shishnag and Kolahoi. Since limestone rock is softer, it gets abraded three times faster than basalt, as confirmed by earth scientists experimentally.[10] This resulted in rapid abrasion followed by valley polishing.

Plucking

Plucking is the process of detachment of rock fragments by the overriding ice. It is carried out by glacier ice along the well jointed, fractured blocks, where rocks are pulled apart mechanically from rock floor.[11] The action is powerful if the plane makes an angle with the surface or if strata are thinly bedded.

Mechanism of Plucking

The major part of glacial erosion involves plucking which weakens rock blocks beneath the glacier. Matthes contended that well jointed rock with favourable spacing of fractures were responsible for plucking, in contrast to tough rocks having few joints. He suggested that joint spacing provides optimum condition for loosening the large blocks of rock from the parent body.[12] He and many other workers succeeded in demonstrating the relationship between rock type and structure on the one hand and depth of glacial excavation on the other.[13, 14]

Process of Plucking

There are two aspects of processes involved in the mechanism of plucking : the provision of fractures in the bedrock and the manner of extraction of loosened block by moving ice (Fig. 35). The origin of bedrock jointing is wholly non-glacial, resulting from tectonics. Apart from jointing being preglacial in origin, there is also the possibility of joints developing contemporaneously with glaciation.[15] Lewis was the first to suggest a link between pressure release and glacial erosion.[16] He argued that normal pressure of ice in thick glaciers could lead the bedrock to cause shallow fractures to develop parallel to the surface after the glacial retreat. Boulton has supported this view and claimed that difference in pressure between upstream and downstream side of the valley may be sufficient to cause rock fractures.[17] Once the jointing and other fractures have been created the way is open for the major part of erosional process to operate. It involves the weakening of rocks by widening them[18] (Fig. 38). There are two schools of thought on the mechanism of plucking. The first recognises the weakening of bedrock prior to glaciation. The lines of weakness develop fractures and joints. These weakened lines are exploited by deep weathering before glaciation.[19, 20, 21, 22] The mechanism of weakening before glaciation is also associated with unloading of land surface due to erosion leading to dilation of joints.[23]

The second school believes that the processes of weakening of bedrock operate beneath the glacier. They suggest that preglacial conditions during the onset of glacial phase can result in the loosening of rock in front of glacier by frost shattering.[24, 25, 26]

In the study region plucking is accomplished in many ways. The major joints in quartzite, basalt, slate and limestone are parallel to preglacial valley slope and are, therefore, preglacial in age. But the mechanism for weakening of the rocks beneath the glacier is due to freeze and thaw process. The Liddar glacier being warm-based glacier, resulted in basal melting of glacier. There was effective supply of meltwater to the base of glacier, which ultimately resulted in deep penetration of water along the joints of the rock. Before freeze and thaw, the meltwater widened and loosened the joints along the joint planes. Later freezing resulted in breaking and loosening of rocks.

The northern parts of the West Liddar valley have well jointed rocks of quartzite, andesite, basalt, slate and calcareous sandstone (Table 3). These rocks are thinly bedded. The quartzites are fairly tough and contain generally one to two metre thick band of greywacke. They have small fractures in the direction of the main glacier. These rocks have been severely plucked and have yielded numerous large erratics. These erratics have been carried to Pahalgam, covering a distance of nearly 25 kilometres. Large erratics have been observed at Liddarwat, Aru, Mundlun, and Pahalgam. Near Liddarwat, a huge erratic with a diameter of 30 metres has been observed. From the lithology of the erratics it has been evaluated that the source region is in Tarsar-Basmai section, where there is a preponderance of quartzite and andesite rocks.

Fig. 38: *A Huge Erratic (A) near Liddarwat Guest House Shows 6 m Wide Fracture (B) in the Centre. Striations Run Parallel to the Fracture*

The rocks unit of the northern part of the East Liddar valley consists primarily of basalt, limestone and calcareous sandstone. These rocks are mostly well-jointed and joints are half to one metre apart. The closely spaced joints are primarily the cause of removal of small and medium fragments and production of adhesive till (Table 3). The examples of the landscape of this type are found near the localities of Shishnag and Zaijpal and they comprise mainly of limestones and slates. The profile of the valley at these sites is deeply rooted, the rotting having resulted primarily from chemical weathering of limestone and partly by the process of plucking and abrasion of slates and limestone.

ABRASION AND PLUCKING

Clean ice cannot abrade any rock. Abrasion is effected only by rock fragments contained in the base of the glacier. The intensity of abrasion is directly proportional to the size of fragments contained in the glacier sole. The fragments are produced by plucking as already discussed. Therefore, the amount of plucking controls the efficiency of abrasion. In the glacial landscape the process of abrasion and plucking operate simultaneously. The glacial erosional features of the study region reveal that abrasive action alone was ineffective for the excavation of cirques and broad troughs at the localities of Shishnag, Zaijpal, Liddarwat, Chandanwari and Aru. They were produced by the combined process of abrasion and plucking (Fig. 35).

GLACIAL EROSIONAL FEATURES

In the glaciated section of the Liddar valley the limited erosional features like whalebacks, roche moutonnees, glacial valleys (troughs) and the cirques were identified and studied in detail in the field (Fig. 39). Each of these features are discussed below with respect to their morphology, distribution and origin.

Striation, Grooves and Polished Surfaces

Striation, grooves and polishing are the indelible marks engraved on the surface of the bedrocks by glaciers. The intensity of the marks depends upon the shape and comparative hardness of abrasion tools. In the polished surfaces there are minute striations and innumerable scratches. These are formed by grains of sand trapped between the abraded boulders in ice and bedrock. The striation consists of lines

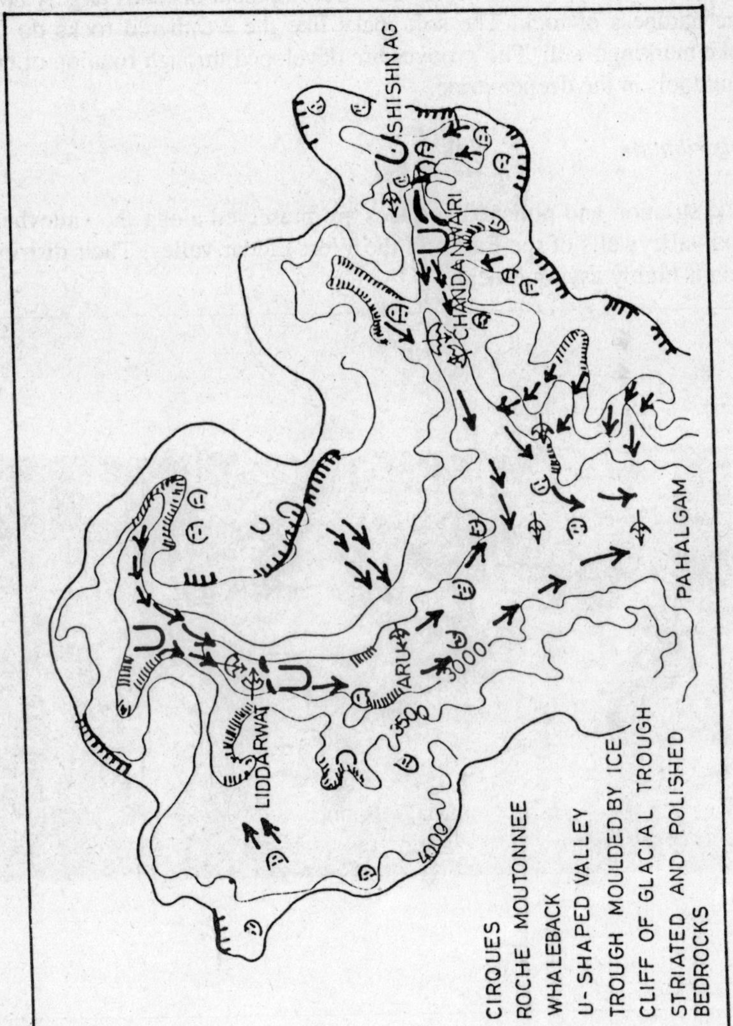

Fig. 39: *Distribution of the Glacial Erosional Features in the Liddar Valley*

usually short in length and engraved on suitable rock surfaces. They are mostly found on inclined valley floor and gently inclined mountain slopes (Fig. 40). The clarity of their development depends largely upon the hardness of rock. The soft rocks like the weathered rocks do not take markings well. The grooves are developed through rotation of cutting tools in the deeper striae.

Distribution

The striation and polished surfaces are preserved along the valleyhead and valleywalls of the East and the West Liddar valley. Their distribution is highly asymmetrical.

Fig. 40A: *Glacially-abraded Valleywall (A) near Pissu Ghati. Striations Run Parallel to Down-stream Direction*

Fig. 40B: *Striated-slates (A) exposed near the Pharsulan till*

East Liddar Valley: Nearly all the striae in this valley are on the volcanics and limestones. Considering the severity of glaciation and abundance of hard rocks which are present in the valleyheads, the number of instances of striated surfaces is comparatively small. The polished surfaces are plentiful in the neighbourhood of Zaijpal basin. The heavily striated and polished surface is prominent at higher altitude near Zaijpal, Dudhal and Wokhbal. The intensity of engraving is maximum in the middle of the valley, moderate towards the glacier snout and low or sparsely striated and polished towards the valley basin of Pahalgam (Fig. 41). The following is the list of striated surface with the main directions of striations (Fig. 42):

1. Surface of volcanics at Manipal (2300 metres above sea level) along the left bank of East Liddar river S 45°W.
2. Surface of limestone on the left bank of river (2400 metres above sea level) S 15°W.

3. Surface of limestone on right bank of river near Nekbattan (2550 metres above sea level) N 28° E and S 08°W.
4. Surface of sandstone at Pharsulan (2600 metres above sea level) S 10°W.
5. Surface of volcanic rock at Dudhal (2850 metres above sea level) S 30°E, S 45°E and S 25°W.
6. Surface of limestone at Chandanwari (2800 metres above sea level) S 45°W and N 36°E.
7. Surface of volcanics near Chandanwari (2950-3000 metres above sea level) S 42°W, S 15°W and N 76°E.
8. Surface of limestone at Bruzut (3000-3100 metres above sea level) S 45°E and S 23°W.
9. Surface of limestone at Shishnag (3500 metres above sea level) S 10°W, S 30°W, S 75°W.
10. Surface of volcanics at Shishnag (3600 to 3700 metres above sea level) N 75°W, N 20°W and N 15°E.

West Liddar Valley: The glacial striae in the West Liddar valley are confined to its northern section, along the series of outcrops of basalt, limestone and slates lying athwart the direction of ice flow. In general, the striae are approximately parallel to valleys in which they occur. The dense siliceous rocks of basalt preserve striae and even polish over vast length of times while limestone and slates loose them quickly. In the steeper slopes of Kolahoi, Tarsar and Basmai the striation and polishing are well preserved. In these sites the basal meltwater has been replacing the smaller fragments continuously. The striated areas is more extensive in comparison to the East Liddar valley. The landscape has been grooved near Liddarwat over a wide area. The polished landscape is comparatively smaller than the East Liddar valley, considering the severity of glaciation and abundance of hard rocks present in the localities of Kolahoi, Basmai and Liddarwat, the number of instances of striated to the exposed nature of valley. The striated surfaces are extensively preserved on basalt, andesite, quartizite and slates and are poorly preserved on limestone and shale. The Liddarwat-Kolahoi section is typically rugged and uneven, the surface is made up of innumerable bosses of rock which are usually striated. The whaleback are located on their upstream ends and sides. The following is the list of striated surfaces along with the main direction of striation (Fig. 42):

1. Surface of basalt at Aru (2600 metres above sea level) along the left bank of West Liddar S 45° to 70°W.
2. Surface of andesite near Aru (2680 metres above sea level) along the left bank of West Liddar S 30° to 45°W.

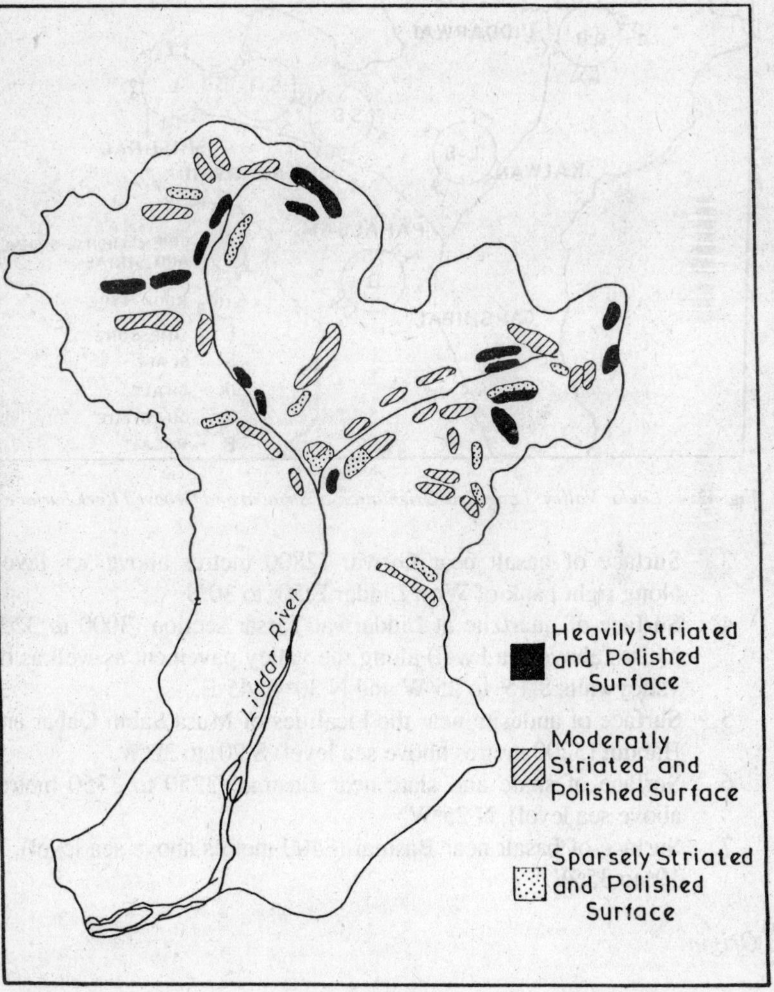

Fig. 41: *Liddar Valley: Intensity of Erosion of Striated and Polished Rock Surfaces*

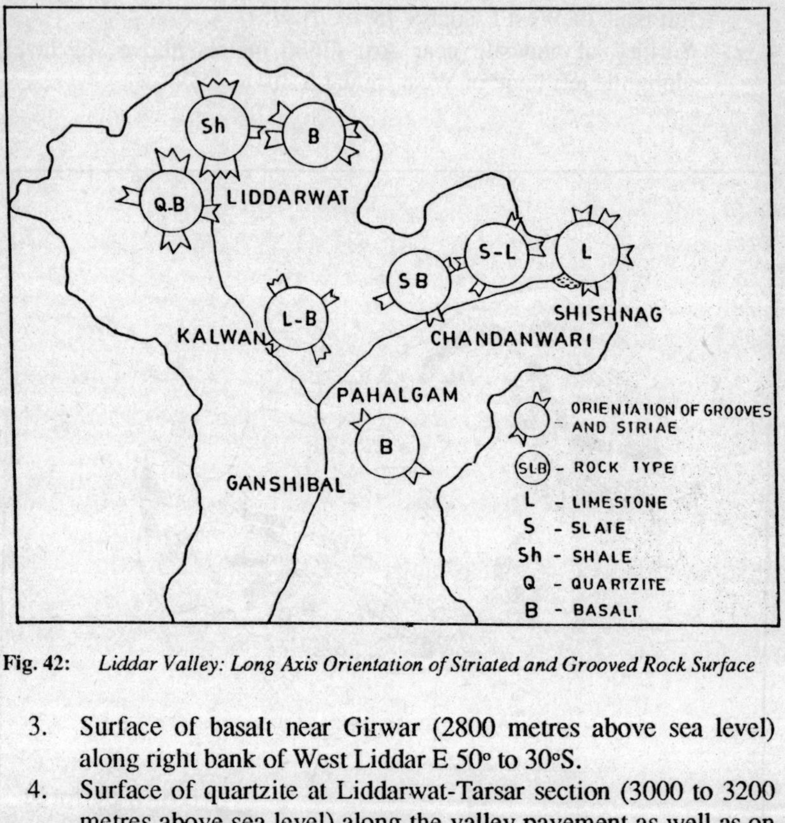

Fig. 42: *Liddar Valley: Long Axis Orientation of Striated and Grooved Rock Surface*

3. Surface of basalt near Girwar (2800 metres above sea level) along right bank of West Liddar E 50° to 30°S.
4. Surface of quartzite at Liddarwat-Tarsar section (3000 to 3200 metres above sea level) along the valley pavement as well as on valleywalls S 15° to 25°W and N 30° to 45°E.
5. Surface of andesite near the localities of Musa Sabin Qabar and Handut (3200 metres above sea level) S 20° to 32°W.
6. Surface of shale and slate near Basmai (3250 to 3350 metres above sea level), N 25°W.
7. Surface of basalt near Basmai (3400 metres above sea level), N 10° to 35°W.

Origin

The distribution of striated and polished rocks in the East and the West Liddar valleys indicates the high concentration of abraded rock surfaces near the middle section of the main valley and along the northern reaches of the hanging valleys (Figs. 39 and 41). In these sections of the valleys there was a continuous supply of abrading material to the sole of the glacier from the higher reaches of the Kolahoi-Basmai and Katsal-Saribal mountains. These mountain cliffs are comprised mostly of closely jointed volcanic, limestone and quartzitic rocks, located on both sides of the West and East Liddar rivers, respectively. The rock fragments of volcanics and limestone abraded the bedrocks of the same lithology in Liddarwat-Kolahoi and Zaijpal-Shishnag section of the West and the East Liddar valleys. It resulted in equal amount of abrasion from each side.[27] Further, there was continuous replacement of small particles by large fragments along the valleyrim of the glacier sole because the melting of ice at the bottom of the glacier was able to remove the smaller particles. The replacement of smaller particles strengthened the capacity of the glacier to hold its engraving tools firmly[28] (Fig. 38). Striation and polishing are, therefore, mainly attributed to basal friction of large angular rock fragments.

The grooves at the Liddarwat site of West Liddar valley are slightly sinuous with soft-flowing outlines of greywacke quartzite. Greywacke being softer was attacked by angular fragments of basalts carried in the base of the glacier. First of all striations were engraved on greywacke beds of quartzite. Later, these striations were gauged deeper due to the rotation of the cutting tool (Fig. 43). The bedrock as well as pavements of Zaijpal-Shishnag section of the East Liddar valley is highly polished. It is composed mainly of limestone rocks. The striation on the pavement is finer. It has resulted in glass like polish. The landscape has been evolved by the sanding action impared by friction of smooth flat surface of boulders frozen firmly into the sole of the glacier by the fine powder derived from rock-flour ground by the ice.[29, 30] This suggests severity of ice erosion, possibly twice that of striation. The edges of the boulders are worn out due to abrasion while forming striation near Zaijpal waterfall. The weight of ice mass could have been maximum over here and supply of abrasion tool of angular fragment could have been small.

Whaleback

Whalebacks are glacially moulded, hard rock surfaces. Their length is greater than their height. They have smooth rock surfaces on all sides which have been produced by glacial abrasion. In spite of the roundish outline, the smooth slope attains angles up to 40°. The length-height ratio in individual whalebacks ranges between 2 : 1 to 4 : 1. The lithology and structure influence their dimension. The mechanics of moving glacial ice plays a dominant role in their formation.

The whalebacks were first recognised in Alps by Saussure who described the rippled, glistering effect produced by glaciers as 'roches moutonnees'.[31] Charlesworth termed them as rock drumlins. They occur as peripheral members of field of drumlin. Linton too suggested the term rock drumlin.[32,] Matthes and others termed such features as whalebacks.[34, 35, 36]

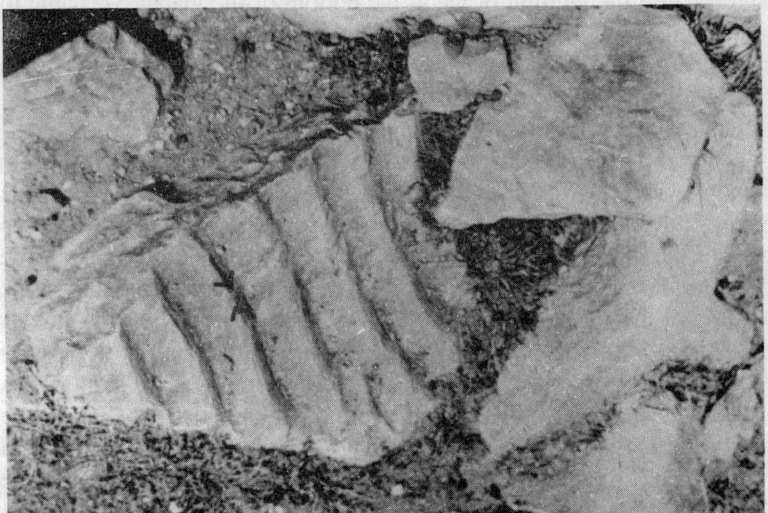

Fig. 43: *Glacial Grooves (A) on a Rock Surface of Liddarwat*

Morphology

Whalebacks in the study region are rounded and elongated hillocks. They are moulded on all sides and are highly striated and polished. They resemble the shells of crocodile[37] (Fig. 44). They have been formed mainly by glacial abrasion and are located in the areas of ice convergence. The individual whaleback is more elongated and its length-height ratio ranges between 3 : 1 to 5 : 1. The long whalebacks are typically located in the East Liddar valley. They are 15 to 30 metres long and 3 to 6 metres high. They have gentle slopes of 3° to 5° (Fig. 45A). The short whalebacks are 5 to 15 metres long and 2 to 6 metres high. The slope ranges between 15° and 30° (Fig. 46A).

Fig. 44A: *Whaleback Field (A) in Zaijpal. Their forms Resemble those of the Shells of Crocodile*

In the East Liddar valley the whalebacks have predominantly tapering long axes and blunted short axes. Their orientation is northeast and northwest respectively. They are primarily composed of hard limestone and sandstones. A whaleback field has been identified near Zaijpal. An individual whaleback looks like a tortoise shell in outline and is spaced 1 to 2 metres apart from each other. Its length-breadth ratio is 1 : 5. In the West Liddar valley the whalebacks are shorter and broader than their counterparts in the East Liddar valley. Their length-breadth ratio is 1 : 2.5 only. They are thick in the middle and tapering towards ends. They are comprised of quartzite and limestone. Near Liddarwat they are formed of quartzite surrounded by soft greywackes.

Distribution

Nearly 30 whalebacks have been identified in the northern part of the Liddar valley of which 24 are located in the East Liddar and the rest in the West Liddar Valley (Fig. 39). In the East Liddar valley 4 whalebacks are located at Shishnag, 18 at Zaijpal and 2 at Dudhal. In Zaijpal 18 whalebacks are clustered in a field of half a square kilometre. They are located along the eastern fringe of the lip of the waterfall. In the West Liddar valley 6 whalebacks have been observed between Satlanjan and Liddarwat at the lip of the riegel.

Fig. 44B: *Smoothed and Polished face of Pahalgam Whaleback. The Arrow Points to the Direction of Ice-flow*

Origin

The morphology of the whalebacks apparently suggest that they were formed in the areas of differential lithology where two to three ice lobes usually converged. The thick ice could pick out the rock weakness to create a landform by two ice lobes, one coming from the Shishram glacier along the East Liddar valley and the other coming from the steep hanging valley of Vaobal. Due to the overriding of the hanging glacier over the main valley glacier, there was lateral movement in the glacier, which resulted in severe erosion along the western side of the valley. It caused irregularity in the bed and produced four whalebacks by sanding action of the glaciers.

On the Zaijpal and Liddarwat sections of the East and the West Liddar valley, the whalebacks were formed by three ice lobes, one coming from main valley glacier and other two coming from steep hanging valleys. The glaciers converged along the narrow steep slope above the present lip of Zaijpal waterfall and Liddarwat riegels. The Zaijpal section is mainly comprised of slates and thin limestone beds. The strike of the beds is 60° northwest, where the movement of the glacier is along the dip of the bed that is northwest. The slates being highly cleaved were eroded away by the main valley glacier of Shishram. The thin beds of limestone present along the eastern margin of the valley section, were attacked by two confluent glaciers. The confluence of the hanging glacier with the main valley glacier resulted in ice radiating laterally towards the east, but maintained their initial

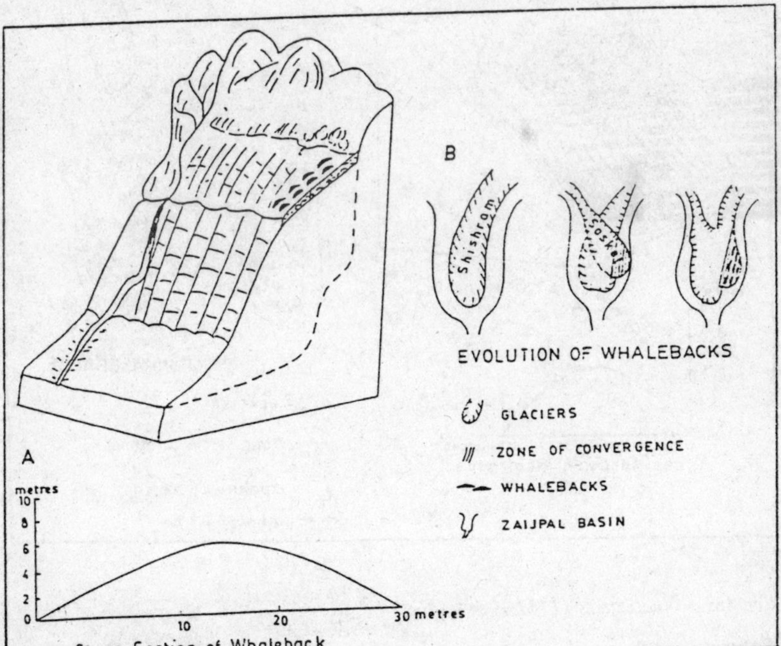

Fig. 45: *Whalebacks: Zaijpal Valley*

direction of movement (Fig. 45B). The lateral glacial extension resulted in thick accumulation of ice followed by severe abrasion of limestone rocks on the bed. It ultimately produced the hard, well-polished whalebacks after the abrasion of softer material. The regular thin bedding-planes of limestone facilitated the formation of whaleback field along the Zaijpal escarpment.

In the Liddarwat section the whalebacks were moulded from hard quartzite rocks along the lip of riegels, by the above-mentioned process. Quartzite being harder than limestone, the whalebacks are higher and steeper than their counterpart in Zaijpal section (Fig. 46B). On the exposed Liddarwat escarpment the accumulation of ice was thin and this resulted in comparatively lesser erosion.

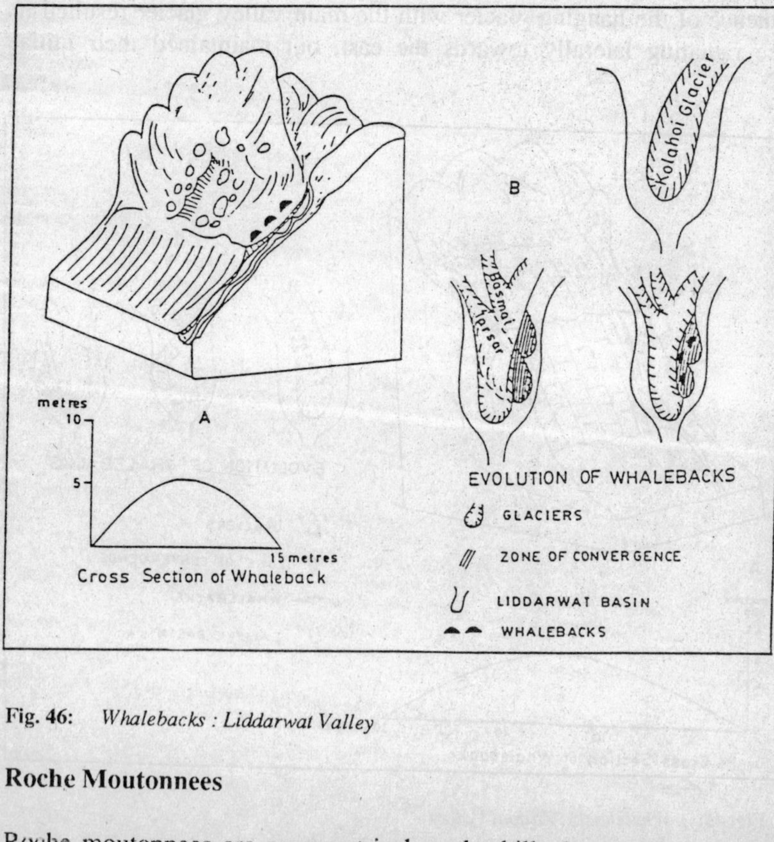

Fig. 46: *Whalebacks : Liddarwat Valley*

Roche Moutonnees

Roche moutonnees are asymmetrical, rocky hillocks produced by the combined action of abrasion and plucking. The abraded and plucked surfaces are well defined. They reveal the thickness of the ice and indicate the side, either stoss or lee as evidenced by striae, that was taken by ice.[38] The roche moutonnees are located at different altitudes of the study region and are observed near the fringe of glacial retreat. The majority of the roche moutonnees are weathered and only a few are preserved from top to bottom (Fig. 47).

Distribution

The distribution of the roche moutonnees is highly selective in the altitudinal zones. They are localised at altitudes of 2150, 2575, 2790 and 3000 metres. They have been identified in the East and the West Liddar valley at Pahalgam, Chandanwari and at Kalwan and Liddarwat respectively (Figs. 39 and 47). The moutonnees of the East Liddar

Fig. 47A: *A Large Roche Moutonnee excavated from quartzitic rock of Liddarwat. Smooth striated summit of stoss side (A) and small cavities along the lee side (B). The direction of ice was from left to right.*

valley are larger than those of the West Liddar. Two large moutonnees have been identified near Chandanwari and one each at the other sites. In the hanging valleys the moutonnees have been observed at Haggan, Dudhal, Tarsar and Girwar.

Morphology

The well developed roche moutonnees formed at Chandanwari and Pharsulan have characteristic parabolic shape. They are located at altitudes of 2575, 2770 and 2800 metres. The moutonnees are 110 to 145 metres long, nearly 90 metres wide and are 40 to 60 metres high from the valley basin. Their stoss-side slope ranges between 30° and 40° and lee-side 60° and 70°. In Chandanwari 2 moutonnees are located adjacent to each other, their polished sides are nearly 80 to 90 metres long. The toe of the moutonnees is located at altitude of 2770 and 2800 metres and the top at 2835 and 2860 metres.

Fig. 47B: *Small Roche Moutonnee near Pahalgam with streamlined stoss-side (A) and quarried lee-side (B). Direction of ice-flow was from right to left.*

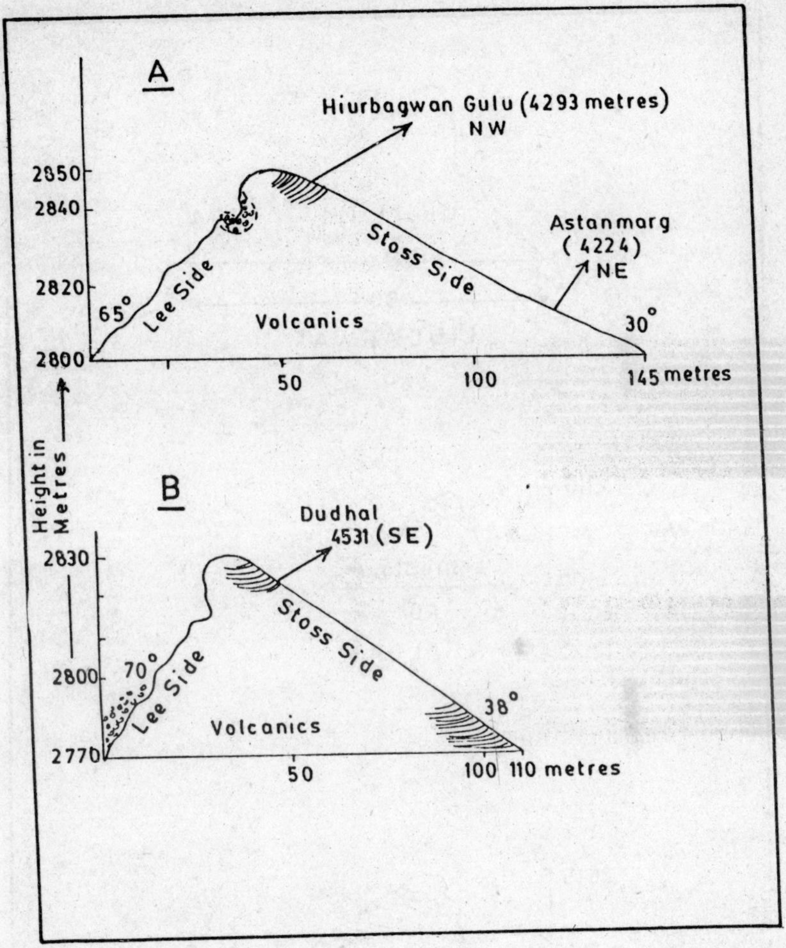

Fig. 48: *Chandanwari Roche Moutonees (A, B)*

On the lee-side of the moutonnees, a huge irregular cavity has been plucked at an altitude of 2830 metres, which is partially filled by talus (Fig. 48). The lee-side slope is steep and shows re-entering angles owing to the partial destruction of knob. The moutonnees are carved from the Panjal volcanics. The top of the moutonnees are well preserved and glacial striae are distinct. In the hanging valleys of the East Liddar the moutonnees have been identified at altitude of 2575 metres at the lip of the riegels near Dudhal and Hagan. The

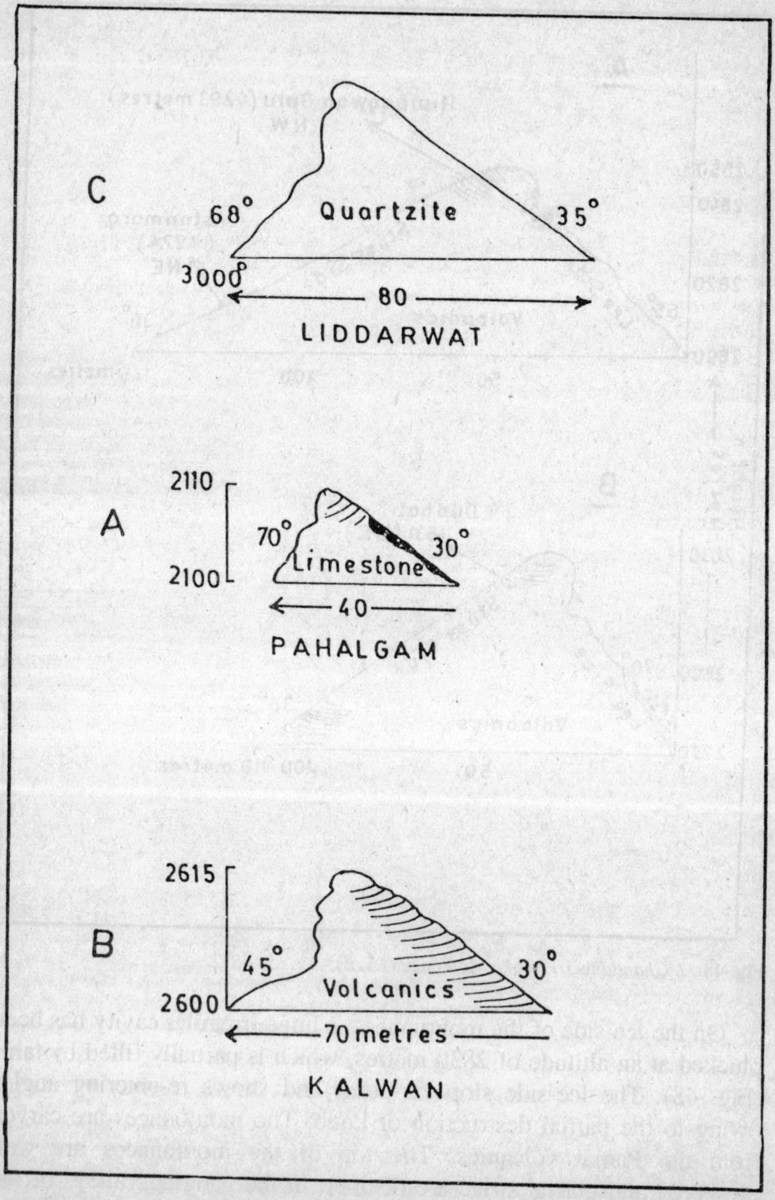

Fig. 49: *West Liddar Valley Roche Moutonnees*

moutonnees are small and are nearly 20 metres high, 75 metres long and 40 metres wide. They are partially destroyed at the top and striae are not traceable. A very small moutonnee has been observed at an altitude of 2100 metres near Pahalgam. It is 40 metres long, 10 metres high and 30 metres wide. The moutonnee has been moulded from limestone and its top is well preserved with faint glacial striae (Fig. 49A).

In the West Liddar valley the moutonnees are 40 to 80 metres long and nearly 20 metres high from the valley basin. At Kalwan and Liddarwat a 70 metres long moutonnees have been cut in the volcanics, at an altitude of 2600 and 3000 metres, respectively. The top of the moutonnees are nearly 15 to 20 metres high and their stoss-sides are nearly 55 metres long. The stoss-side slopes have angles of 25° and 35° and lee-side of 65° and 75°. The leeward slopes are highly irregular from top to bottom and talus are deposited near the base (Fig. 49B and C). In the hanging valleys the moutonnees have been observed near Girwar and Tarsar lake. They are primarily cut in quartzite and are highly weathered, but the tops of moutonnees are still intact at Girwar and are marked with striation.

Origin

The distribution of the moutonnees reveal that they have unusually large size at Chandanwari, which sometimes compel the observer to attribute their origin to preglacial landform processes. But by observing the striae near the moutonnees head one can deduce their origin from glacial action. The moutonnees have been formed by combined action of three different ice masses, coming from Astanmarg (4224 metres), Hiurbagwan Gulu (4293) and from the main East Liddar valley, as indicated by the orientation of striation (Fig. 48). The second moutonnee formed at an altitude of 2770 metres has been formed mainly by Dudhal glacier and the secondary by the East Liddar glacier (Fig. 48B). At Pahalgam the subdued moutonnee top has faint striae lines in the northeast direction. This indicates the direction of the ice movement. It also suggests that the moutonnees have been formed primarily by the East Liddar glacier. Similarly at Girwar and Liddarwat, in the West Liddar, moutonnees have been moulded by two ice masses coming from the main valley glacier of the West Liddar and the hanging glaciers of Tarsar and Girwar.

The actual shape of the moutonnee reflect their origin to abrasion and plucking. The upstream side in most cases is streamlined with striae, suggesting abrasion. The lee-side is subjected to block loosening and removal rather than to abrasion. The smoothed and striated bedrocks near the summit of the stoss-side of the moutonnees of Chandanwari, Pahalgam and Liddarwat reveal that there had been overspilling of ice. The overspilled ice slipped along the leeward side and attacked the volcanics and quartzitic bedrocks along their joints and fractures. This resulted in block loosening and enlargement of small cavities. In the Chandanwari moutonnee the loosened block removed from the top of the lee-side has fallen in the floor of large cavity. At Pahalgam the loosened blocks have fallen on the base of the moutonnees because the lee-side is very steep and broken (Figs. 48 and 49).

Glaciated Valleys

The glaciated valleys of the study region are asymmetric and approximately parabolic in form. From the field evidences and by comparing with non-glaciated valleys of the South Liddar, one infers that the rounded forms from Kolahoi to Pahalgam and from Shishram to Ganshibal which characterise the upland region presently were formed by the Pleistocene glaciers. The valleys were generally V-shaped in cross-section with the characteristic series of interlocking spurs projecting on to the stream. During the Pleistocene period three stages of glaciation were recorded as discussed in the nineth chapter. The evidence of the first two stages was destroyed by the subsequent glaciation. During the third glaciation the ice tongues extended up to Ganshibal. Then followed a period of retreat. As a result of glaciation the form of the East and the West Liddar valleys and their tributaries were altered. The V-shaped valleys became broader and displayed irregular U-shaped trough forms from the valleyheads of Liddarwat and Shishnag to Aru and Chandanwari respectively (Fig. 39). The upper end of troughs were further deepened by more than 100 metres and the lower end by 200 metres. The valleys widened by nearly 400 metres in the upper part, 1000 metres in the middle part and 2000 metres in the lower part. The tributary valleys were cut deeper to form hanging valleys which joined the main valley of the East and the West Liddar at steep angle.

A V-shaped notch has been cut at the base of the U-shaped valley near Pahalgam. In the upper and middle sections of the trough, the valley has developed valley-in-valley profile. This suggests that the present Liddar valley is actually a parasite on the preglacial V-shaped valley. The V-shaped valleys have been identified at the base of the U-shaped valleys at Chandanwari, Pharsulan, Aru, Mundulun and Pahalgam. Their presence is by no means unusual; such distribution has also been observed in Swiss Alps and Norway.[40, 41 and 42]

Morphology

The troughheads of the East and the West Liddar are primarily moulded from volcanics. The side walls of these troughs are very steep and wide (Fig. 50). About 2 kilometres downstream from the troughhead at Shishnag and Satlanjan, the valley trough broadness as it passes over limestone and shales (Figs. 51 and 52). About 3 kilometres downstream from the same point near Zaijpal and Liddarwat, the profile becomes gentle and more steeply sloping as it crosses beds of slates, limestones and sandstones of the East and the West Liddar respectively. At these sites the trough floor narrows and is superimposed on the chain of riegels produced between the waterfalls. The valley floor drops by 15 to 50 metres. Further 2 kilometres downstream the trough broadens symmetrically at Zaijpal and Liddarwat and assumes and U-shaped form (Figs. 53C and 54A). Below Zaijpal and the Liddarwat the rock basins are cut in the limestone beds after the erosion of the overlying volcanic beds. At these sites the valleys narrow. Steep risers have been produced along the volcanic beds. Further South and East and the West Liddar valley widen asymmetrically at Chandanwari and Aru, respectively. This section of the valley is comprised of micaceous sandstone and calcareous slates. Four kilometres downstream towards Pahalgam the valley has gentle gradient (3°-5°), but at Mundlun and Pharsulan the slope once again gradually increases near the interlocking spurs projecting on to the U-shaped valley. A deep V-shaped notch has been produced by Liddar river.

Longitudinal Profiles

Glaciation has caused change in the form of the longitudinal profiles of the East and the West Liddar valleys and their hanging valleys. In this change the character and distribution of joints in the bedrock has played an important role. The variations in the profiles of the main valley are minor but in the hanging valleys it is appreciable.

Fig. 50: *A broad U-shaped Valley (A) at Liddarwat, Showing Glacial Trough Carved below Higher Level Bench (B) of Volcanics*

East Liddar Valley: The troughhead of the East Liddar valley is 1.8 kilometres long and irregular in form. A more varied rock structure has given rise to irregular form but overall the troughhead has convex form, largely because it is joined by three hanging valleys in the east. The entire valley stretches over a length of 24 kilometres from the troughhead. The profiles of the main and hanging valleys indicate characteristic change from troughhead to trough terminus (Fig. 51). The profiles aligned with the middle of the trough, are concave in form. The long profile of the East Liddar reveals that the floor of the valley is broad stairway with three series of treads having lengths of 4, 6 and 14 kilometres from northeast to southwest respectively. The treads are separated by 250 and 100 metres high stairway, situated at elevation of 3600 and 3000 metres respectively. Inbetween the two stairways two rock basins have been cut in the profile. The higher basin has developed into lake.

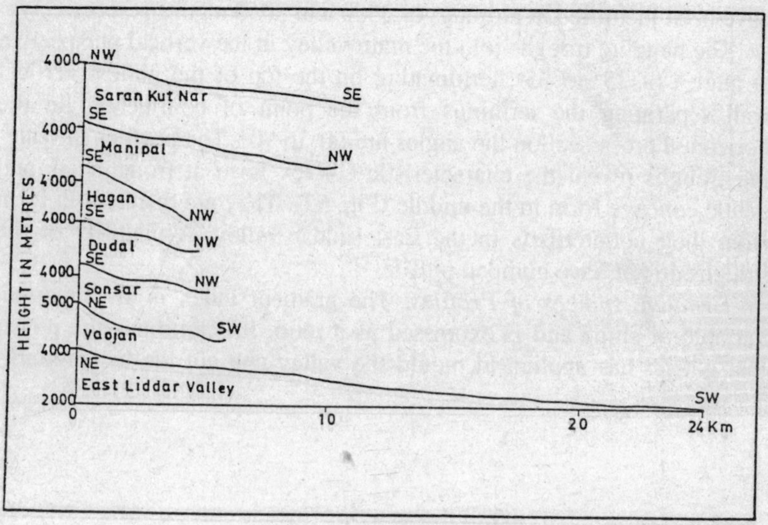

Fig. 51: *East Liddar Glacial Trough and its Hanging Valleys*

The hanging troughs join the main valley at a slope of 35° to 50°. A majority of the hanging troughs have developed convex form near the upper and the terminus end of the profiles. The profiles aligned with the middle of the trough are concave in form (Fig. 51). The convex form near the terminus looks like a hump. The profile of the Haggan hanging valley is entirely different from those of the rest. It is steep and narrow with convex form near the terminus. The hanging troughs terminate at higher altitudes above the main valley (Fig. 51).

West Liddar Valley: The glaciated section of the West Liddar valley is 31.5 kilometres long. The valleyhead is located near the Kolahoi and its mouth near Pahalgam. The troughhead is nearly 1.25 kilometre long, 1500 metres in amplitude and is convex in form. The long profile of the valley is gentle and comparatively less steep along the stretch of valleyhead and the waterfall. The floor of the valley is a broad stairway with three series of treads separated one from the other by 300 and 200 metres high stairways (Fig. 52). The three series of treads are 3, 11 and 17.5 kilometres long and their stairways are located at altitudes of 3600 and 3000 metres respectively. Between the stairways two rock basins have been identified at altitudes of 3550 and 3050 metres respectively.

The hanging troughs join the main valley in the vertical perspective at angles of 25° to 35°, terminating on the top of the almost vertical wall separating the terminus from the point of confluence. In the horizontal cross-section the angles are 60° to 90°. The profiles of hanging troughs reveal the characteristic convex form at troughhead, and gentle concave form in the middle (Fig. 52). They are different in form from their counterparts in the East Liddar valley. A majority of the troughs do not have humped profile.

Gradient Indices of Profiles: The gradient index is an important variable of slope and is expressed as a ratio. It determines the power that glacier has applied to mould the valley and cut the bench-marks

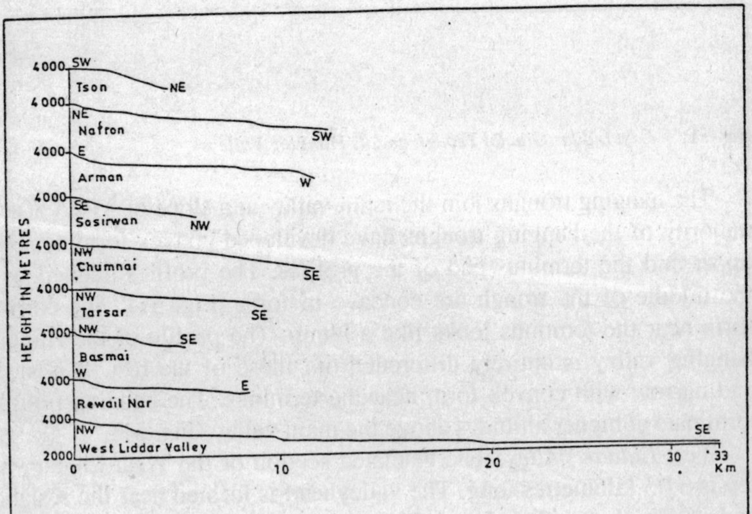

Fig. 52 *West Liddar Glacial Trough and its Hanging Valleys*

and the rock basins on the bed. The shape and altitude of the valley are related to the lithology of the beds at different sections of the valley located at different altitudes. The longitudinal profiles of the East and the West Liddar valley were analysed for determining the gradient indices at different sections. They were calculated by using the formulae given by Hack – Seeber and Gornitz.[43, 44] They applied two indices to the selected Himalayan rivers. A majority of the Himalayan rivers are glacier-fed and their main course in the mountainous region is primarily through glaciated valleys. Therefore the gradient index is more applicable to glaciated valleys.

The gradient index of the East and the West Liddar valleys have been calculated from the following formula given by Hack.[45]

$$K = \frac{H_i - H_j}{\ln L_j - \ln L_i}$$

Where

K = Slope of the valley profile termed gradient index.

H_i = the highest elevation of the troughhead of the valley near the snout of the glacier (3800 metres).

H_j = the height of the valley at the entrance to the fluvial section 2100 metres).

L_i = the length of the glaciated valley above an altitude of 3800 metres. For the East Liddar valley it is 1.25 kilometre and for the West Liddar valley 1.20 kilometre.

L_j = the East Liddar glaciated valley is 24.5 kilometre long and the West 31.5 kilometres.

The valleys extend from an altitude of 2100 to 3800 metres. The gradient index for the East Liddar valley is:

$$K = \frac{3800 - 2100}{\ln 24.5\text{x}10^3 - \ln 1.25\text{x}10^3} = 571.327$$

and for the West Liddar valley is:

$$K = \frac{3800 - 2100}{\ln 31.5\text{x}10^3 - \ln 1.20\text{x}10^3} = 526.83$$

The gradient index of small reach (SL) for the different segments between the consecutive contours has been calculated for the East and the West Liddar valley from the Seeber and Gornitz formula.[46]

$$SL = \frac{\Delta H \times L}{\Delta L}$$

where

SL = gradient index of short reach.

ΔH = the elevation difference between the upper and lower parts of the section. It is calculated from the contour lines and from the field measurements.

ΔL = Planimetric length of the given valley section.

L = Length of the section of the valley from the troughhead to the middle point of the section.

The gradient index of short reaches (*SL*) of different sections of the East and the West Liddar valley were reduced to relative ratio with respect to the slope of the valley (*K*), by dividing *SL* by *K*. A ratio of 1.0 indicates that the gradient of a given section (tread) is the same as that of the logarithmic profile of the valley. A small ratio of less than 1.0 indicates a gentle section and a large ratio of more than 1 indicates a steeper section (tread). A sharp increase in the value indicates a bench cut or stairway in the valley bed (Figs. 51 and 52). In the East Liddar valley *SL/K* values range between 0.204 and 6.826, whereas in the West Liddar valley the values range between 0.189 to 5.093 (Table 17). Further the values show a sharp rise from 2.05 to 6.826 and 2.61 to 5.093 at altitudes of 3500–3550 and 2800–3000 metres in the East and the West Liddar valley respectively (Tables 17 and 18) (Figs. 51 and 52). The values indicate that the sections (treads) are significantly steeper ($SL/K \geq 2$) and much steeper ($SL/K \geq 5$) than the ideal profile.

Cross Profiles: The transverse profiles at different sections of the East and the West Liddar valleys reveal the cross-sectional shape of valley. The shape of the valley is a product of variations in the rates of vertical lowering and lateral valley cutting by glacial erosion processes.[45] The processes lead to the deepening and widening of the East and West Liddar valleys. The cross-section shape of the valley is therefore expressed by a ratio of depth and width.[47] The depth-width ratio at the valley top of the East and the West Liddar ranges between

0.31 and 0.46 and 0.33 and 0.42 respectively (Figs. 53 and 54). It suggests that the East Liddar valley troughs are comparatively deeper than their counterpart. The profiles show the ideal U-shaped form near Vaojan-Zamtirnag and Kolahoi-Satlanjan sites of the East and the West Liddar respectively. The side walls of the troughs are very steep and deep. The trough top widths of Vaojan-Zamtirnag and Kolahoi-Satlanjan range between 750–1100 and 1100–1600 metres, respectively, and their depths 500–1000 and 400–800 metres, respectively. The valley is somewhat hemispherical in shape. The troughs at Zaijpal and Liddarwat widen to 2100 and 2500 metres respectively and their depth decreases to 800 and 825 metres, respectively. The U-shaped outline of the trough are still maintained but they become broader and shallower. South of Zaijpal and Liddarwat the troughs become wider and shallower and their U-form distorts. Near Pahalgam the trough becomes highly distorted and a V-shaped notch has been observed at its base. The cross profiles from the valleyhead of the East and the West Liddar to the Zaijpal and the Aru reveal a valley-in-valley topography (Figs. 53C and 54D).

Table 17A: *Gradient Indices of Different Segments of the East Liddar Glacial Valley*

Segment	*Length of segment (L) (metres)*	*Relative relief of segment (ΔH) (metres)*	*Planimetric length of valley section (ΔL) (metres)*	*Gradient index of segment* $SL = \frac{\Delta H \times L}{\Delta L}$ *(metres)*	*Gradient index of the valley (K) (metres)*	*Relative ratio of slope (SL/K)*
1.	625	400	1250	200	571	0.35
2.	1500	50	500	150	571	0.26
3.	1875	100	250	750	571	1.32
4.	3500	100	3000	117	571	0.21
5.	5500	150	1000	825	571	1.44
6.	7750	200	3500	443	571	0.78
7.	9750	200	500	3900	571	6.83
8.	10250	50	500	1025	571	1.80
9.	12000	100	1000	1200	571	2.05
10.	13500	150	4000	506	571	0.89
11.	16750	150	2500	1005	571	1.75
12.	21250	300	6500	981	571	1.71

Source: Calculated from the Longitudinal Profile of the East Liddar Valley.

Table 17B: *Gradient Indices of Different Segments of the West Liddar Glacial Valley*

Segment	*Length of segment (L) (metres)*	*Relative relief of segment (ΔH) (metres)*	*Planimetric length of valley section (ΔL) (metres)*	*Gradient index of segment* $SL = \frac{\Delta H \times L}{\Delta L}$ *(metres)*	*Gradient index of the valley (K) (metres)*	*Relative ratio of slope (SL/K)*
1.	625	200	1250	100	527	0.19
2.	1875	200	1250	300	527	0.57
3.	2750	250	500	1375	527	2.61
4.	6375	350	6750	330	527	0.62
5.	10312	150	1125	1375	527	2.60
6.	15187	250	8375	453	527	0.86
7.	20125	200	1500	2683	527	5.09
8.	26625	300	11500	694	527	1.32

Source: Calculated from the Longitudinal Profile of the West Liddar Valley.

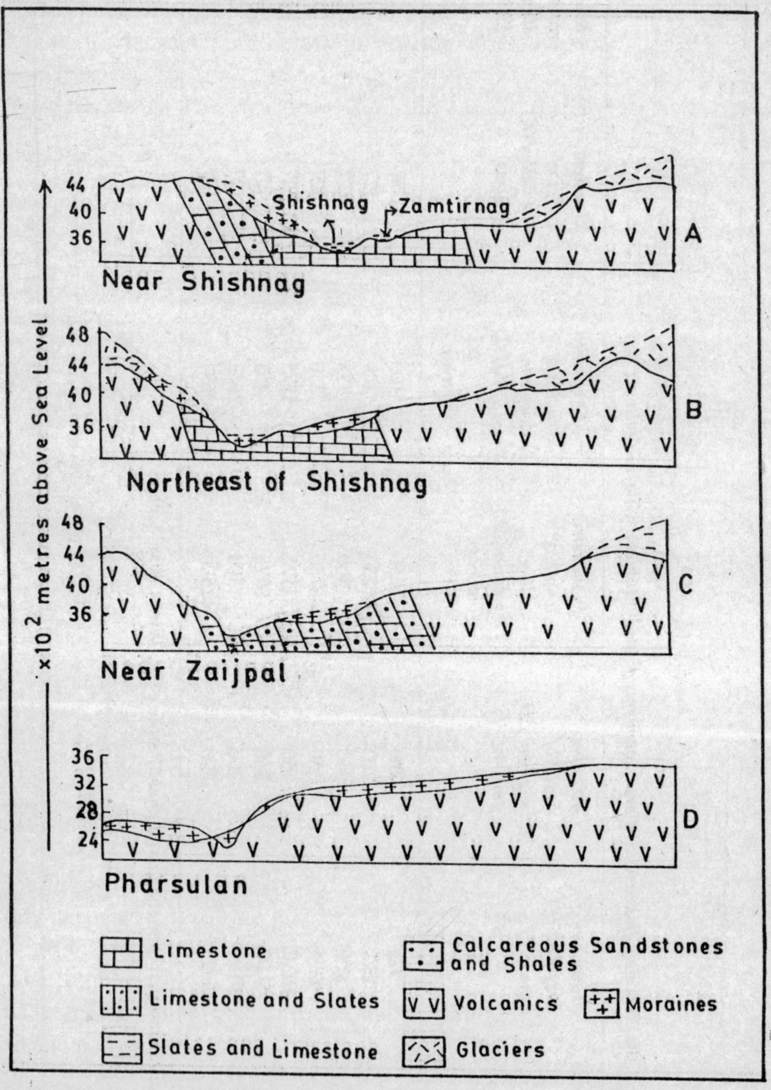

Fig. 53: *Cross Profiles with Geological Cross-sections of East Liddar Glacial Valley*

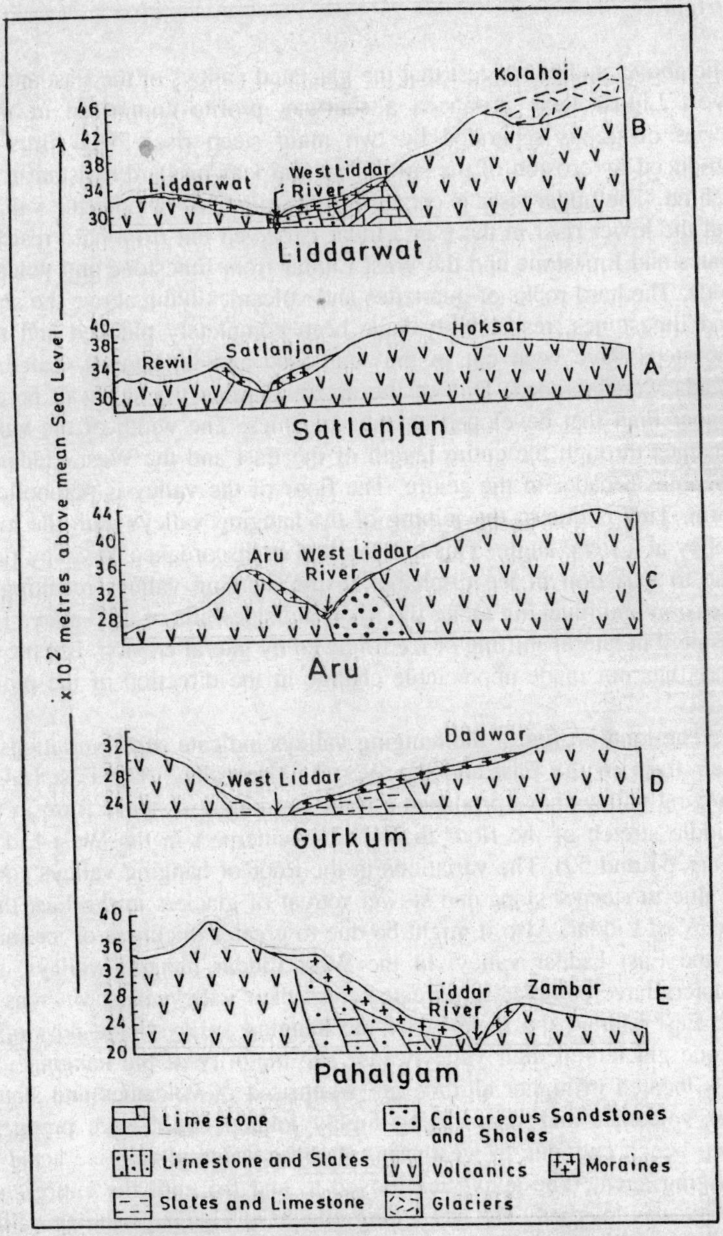

Fig. 54: *Cross Profiles with Geological Cross-sections of West Liddar Glacial Valley*

Origin

The above studies suggest that the glaciated valleys of the East and the West Liddar have produced a stairway profile comprised of three series of treads separated by two main steep rises. The risers are produced by erosion of the softer material leaving hard resistant rocks behind. The upper riser is carved in volcanic beds of both the valleys, but the lower riser in the East Liddar is carved out from hard resistant slates and limestone and the West Liddar from limestone and volcanic beds. The hard rocks of quartzites and volcanics lying above the slates and limestones, respectively have been completely plucked and thus the steps have been cut in the sandstone and limestones. Later the treads were abraded. Due to the above reasons the stairway here is higher than that developed on the volcanics. The width of the valley changes through the entire length of the East and the West Liddar. It becomes broader in the centre. The floor of the valley is parabolic in form. This is due to the joining of the hanging valleys with the main valley at a steep angle. This has resulted in discordance of valley floor due to variation in ice discharge in the hanging valleys resulting in abrasion and plucking along the floor and sidewalls of the valley. This resulted in lateral shifting of ice followed by lateral erosion. But the ice mass has not made appreciable change in the direction of the movement.

The long profile of the hanging valleys indicate some variations in their form in the East and the West Liddar valley. The East Liddar hanging valleys have produced a more prominent concave form in the middle stretch of the floor than their counterpart in the West Liddar (Figs. 51 and 52). The variations in the floor of hanging valleys could be due to steeper slope and slower retreat of glaciers in the East than the West Liddar. Also it might be due to greater thickness of ice mass in the East Liddar valley. In the West Liddar hanging valleys, the glaciers have completely retreated from their valleyheads, whereas in the East Liddar, the majority of the hanging valleys have developed cirque glaciers at their valleyheads. The majority of the hanging valleys located in higher altitude are comprised of volcanics and slates. The volcanics and slates being highly jointed could have produced deep bench cuts due to weathering. During accumulation ice acted as gouging agent. The active ice moved to and fro until the entire concavity was lowered. The steep long profile of Hangan hanging valley could have been formed by landslide which destroyed all irregularities in the glacial profile except the one present near the valleyhead. A steep scarp has been identified at an altitude of 3600 metres along the volcanic traps. The humped form in longitudinal profiles of different hanging valleys of the East Liddar reveals that there could have been a reversal in the floor of the glacier during the three phases of advance and retreat.[48]

Cirque

The cirques are universal in occurrence in the glaciated mountain regions and the most recurrent worldwide feature in glacial landscape. They are confined to the areas of the present or former glaciation. As a result of this broad distribution they have received numerous names, such as corries or cwm.[49, 50] The cirques are the most recurrent geographic features in the glacial erosional landscape of the Liddar valley. They are distributed between the heights of 2200 and 4100 metres. Most of the cirques lying above 3800 metres are occupied by glacial ice. These are called cirque glaciers. There are nearly 17 cirque glaciers ranging in length between 0.1 and 6 kilometres. The ice free cirques lie between 2000 and 3800 metres, with mean elevation of 2975 metres. About 65 per cent of the cirques are located between the altitudes of 2500 and 3600 metres, 20 per cent above the 3600 metres and only 15 per cent lie below 2500 metres.

Considerable work had been accomplished in the morphology and distribution of cirques in the Alps and Cordillera.[51, 52, 53] Very little work has been conducted in the Himalaya and, in particular, in the study region. Cirques have been identified in the Himalaya at different levels below the present snowline. On the basis of cirque levels two to three Pleistocene advances of the Himalayan glaciers have been proposed.[54, 55, 56] The present study deals with the distribution, orientation and morphometry of the cirques, and is a preliminary contribution to the same.

Methodology of Analysis

Eighty-one cirques were investigated. A majority of ice free cirques were identified during the field investigation. Some of the inaccessible cirques, mostly cirque glaciers, were identified on 1 : 50,000 topographic sheets and satellite imageries. A cirque was identified as hollow, bounded upstream by steep cliff, but open downstream (Figs

55 and 56). The break of slope between the cirque floor and mountain slope was precisely located on topographic sheets. The various geometric dimensions pertaining to the length, height and breadth were measured for individual cirque.[57] The lip height was measured by altimetre and orientation of cirque floor by compass. The height of the cirque was measured from the crest of the lip and breadth was measured perpendicular to the long axis (Fig. 55). For the inaccessible cirques all these variables were measured from the contours, spot heights of topographic sheet and long profiles of each cirque.

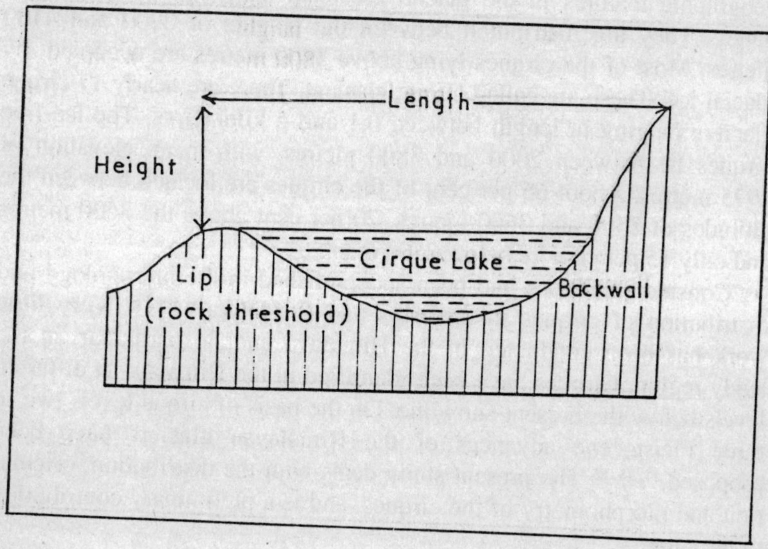

Fig. 55: *Cross-section View of Cirqu*

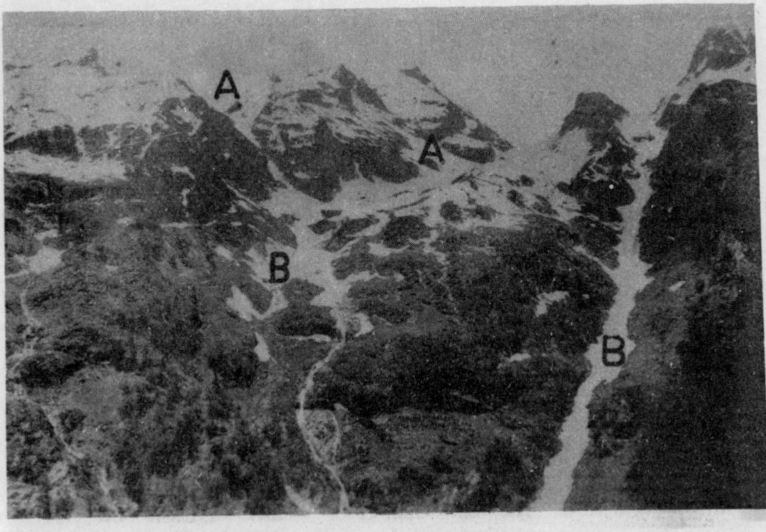

Fig. 56A: *A Small Cirque Glacier (A) near the Trough Head of Graîbal Mountain of Shishnag. Little Ice now Remains Along the Chute Slopes (B) Beneath the Present Day Winter Snow*

Distribution

For description and analysis of the field area was divided into four distinct mountain groups — Kolahoi, Shishram, East Liddar and West Liddar mountain ranges. The first three mountain ranges are encircled by snow traps. In each of the mountain group the icefree cirque length varies considerably between 50 and 2000 metres, but distinctly falls into three groups with mean length of 155, 478 and 1040 metres. The three groups of cirques identified on the basis of length are: small cirques (50 to 200 metres), medium cirques (300-700 metres) and large cirques (800-2000 metres).

Kolahoi Mountain Group: Eighteen cirques have been identified in Kolahoi mountain ranges, out of which 3 have developed cirque lakes (tarns) (Fig. 57A) (Table 18). The icefree cirques are confined within a narrow altitudinal zone ranging from 3250 to 3800 metres. The cirque

Fig. 56B: *A Cirque Stairway (A) near Kolahoi Ganj area. Moraine Ridge (B) Extending from the Lowest Cirque Lip*

glaciers are located above the altitude of 3800 metres. The 4 cirque glaciers have been identified in this group, the largest being Kolahoi (6000 metres long) and the smallest the Hoksar (600 metres long).

There are 6 small and 12 medium size cirques in Kolahoi mountain ranges. The small cirques are located at higher altitude ranging between 3250 and 3950 metres (above sea level). The medium cirques are located at lower altitude ranging between 3250 and 3700 metres (above sea level). As many as 72.3 per cent of the total cirques are located at the mean altitude of 3508 metres (Table 22). All of them have south-north orientation (Table 23). The remaining 27.3 per cent of cirques are located at mean elevation of 3730 metres and have north-south orientation (Fig. 58).

The small cirques are cut in the basaltic and quartzitic rocks of Kolahoi, Musa Sabin Qabar and Basmai anticline. Their headwalls are well developed and are steep. They exhibit strong lithological and structural control. The cirques are very deep and have been formed into the limestone with the overlying Panjal trap getting eroded. They look very fresh and their floors are covered by rock scree and fragments. The bottom of the headwall and cirque floor are well polished. Small moraines extend from the lip of the cirque (Fig. 56B). The medium cirques are cut in shale and limestone. They are shallow and elongated. Their headwalls are usually gentle (3°-7°). Near Hoksar 2 cirques were identified in shales. They have well developed cirque floor but uneven backwall. The cirque floors have been carved in the shale and at the threshold of the floor there is an absence of well defined lip.

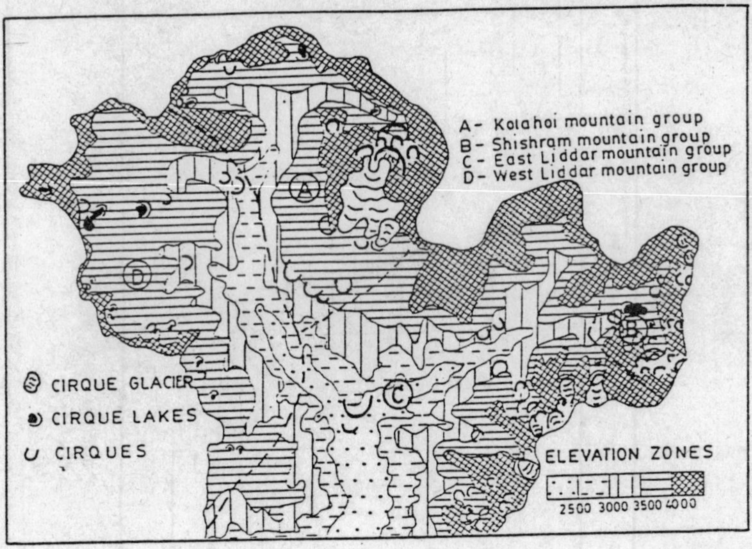

Fig. 57: *Liddar Valley Cirque Distribution*

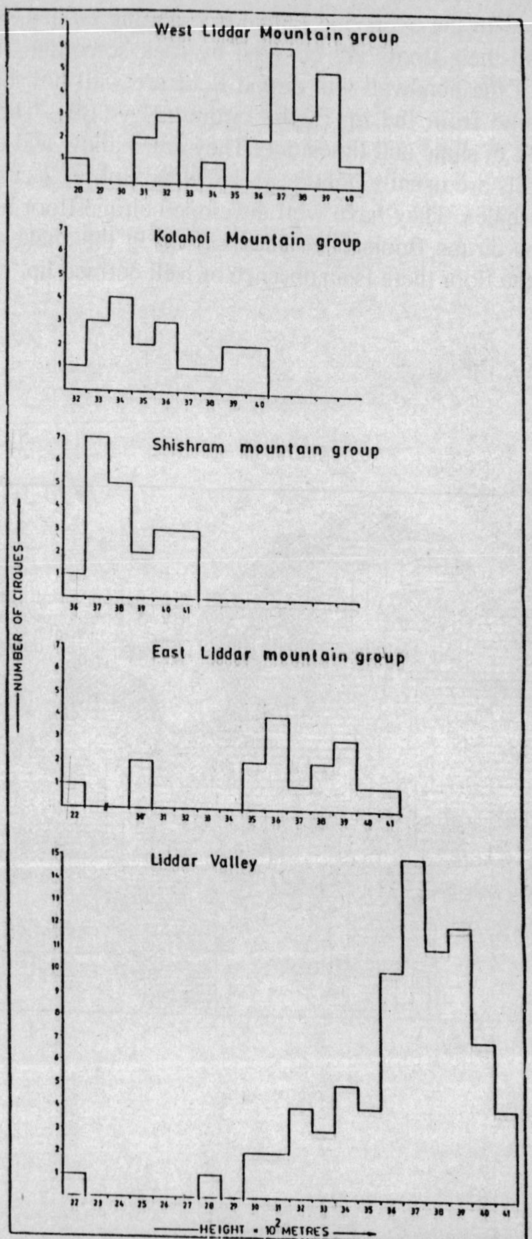

Fig. 58: *Cirque Distribution Lip Height*

Table 18: *Correlation between Cirque Variables of the Kolahoi Mountain Group*

Sl. No.	*Station*	*Direction of cirque floor*	*Elevation of cirque lip (metres, above sea level)*	*Length of cirque floor (metres)*	*Breadth of cirque floor (metres)*	*Height of the cirque from cirque lip (metres)*	*Ratio between length and height*	*Ratio between length and breadth*
1.	Kolahoi East	330°	3800	500	247	190	2.6	2.1
2.	Kolahoi East	345°	3940	195	45	20	9.8	4.3
3.	Cirque Lake	115°	3600	450	100	125	3.6	4.5
4.	Cirque Lake	332°	3400	300	90	150	2.0	3.3
5.	Kolahoi West	335°	3900	150	78	50	3.0	1.9
6.	Cirque Glacier	340°	4000	6000	2100	1800	3.3	2.8
7.	Dudnag	140°	3700	400	250	180	2.2	1.6
8.	Dudnag	342°	3600	400	100	125	3.2	4.0
9.	Dudnag	305°	3500	200	400	50	4.0	0.5
10.	Cirque Lake	310°	3310	450	125	130	3.5	3.6
11.	Hoksar	322°	3400	1200	425	340	3.5	2.8
12.	Hoksar	10°	3520	325	105	95	3.4	3.1
13.	Hoksar	18°	3400	500	110	200	2.5	4.5
14.	Musa Sabin Qabar (Cirque Glacier)	245°	3900	1275	605	630	2.0	2.1
15.	South Kolahoi	186°	3840	600	192	232	2.6	3.1
16.	South Kolahoi	210°	3600	245	100	120	2.0	2.5
17.	South Kolahoi	45°	3400	300	110	140	2.1	2.7
18.	South Kolahoi	15°	3250	220	95	52	4.2	2.3
	MEAN		3608	761.7	293.2	257	3.3	2.1

Source: Field Observations.

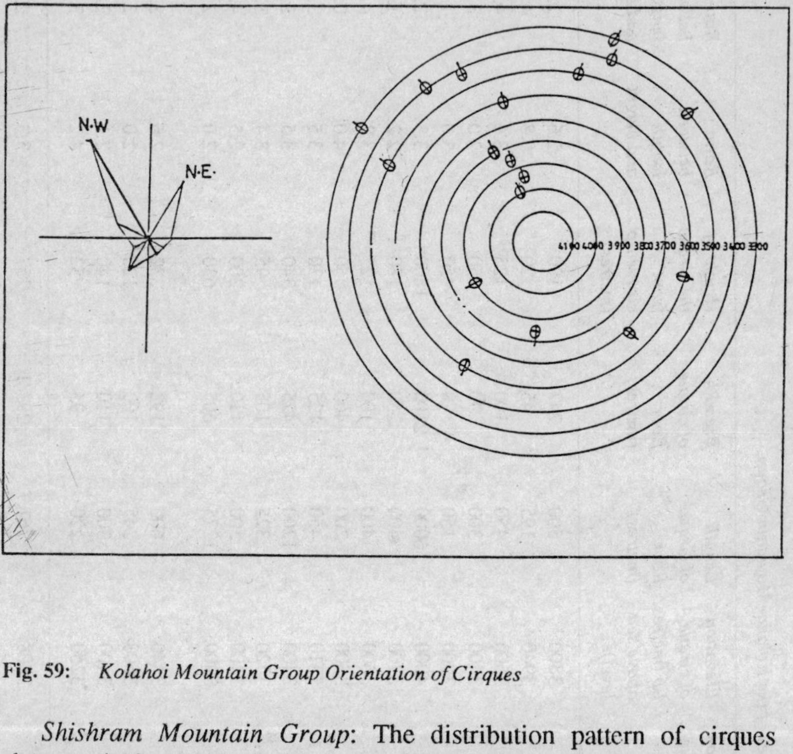

Fig. 59: *Kolahoi Mountain Group Orientation of Cirques*

Shishram Mountain Group: The distribution pattern of cirques show typical gradation with the smallest of them located at higher altitude (4120-3720 metres), the largest in the middle (3500-3600 metres) and the medium in the lower altitudes (below 3800 metres). In this orographic group 23 cirques have been identified, of which 8 are small, 10 are of medium size and 5 are large cirques (Table 19). The icefree cirques are located between the altitudes of 3550 and 3800 metres. The distribution shows radial pattern, with the largest cirques of Shishnag in the centre. The 8 cirque glaciers are located above the altitude of 3800 metres. They cluster around the Shishnag valley in semicircular fashion (Fig. 57B). Six small cirque glaciers lie on the northwest and 2 large cirque glaciers on the northeast; the largest being Shishram cirque glacier (5250 metres long) and the smallest the Gratbal (50 metres long). The smallest of the cirques are located on the higher segments of slopes of Gratbal, Gulol Gali and Mahagunus. They lie west of the eastern ice fence. About 85.7 per cent of the total cirques are located at an altitude of about 3820 metres (Fig. 58). They have south-north orientation (Table 23).

The small cirques are carved in andesitic rocks of Mahagunus, Gulol Gali and Gratbal. They are deep and well developed. The cirque lips are developed 5 to 10 metres high from the sloping valleys. They are highly polished. The headwalls of the cirques are weathered near Gratbal. Most of the icefree small cirques are covered with angular rock fragments. The cirque floors are striated. The medium cirques are mostly angular in outline. Nearly 70 per cent of them are located below an elevation of 3750 metres (Table 22). The cirques are cut in limestone and slates. The cirque floor and cirque lips are well developed, but their headwalls are unevenly developed. The medium cirques near Shishnag, Zamtirnag and Konnag are formed across the strike of the limestone rocks. The limestones trend in northwest to southeast direction and are moderately jointed. The prominent direction is along the bedding plane. About 71 per cent of cirques within the area have northwesterly component (Table 23) (Fig. 60). This clearly reveals that there is a direct relationship between the strike of limestone bed and cirque orientation. Two well developed cirque stairways have been observed in northeast and southeast of Shishnag. The cirque stairways to the northeast of Shishnag have been identified at altitudes of 3680 and 3750 metres. The higher cirque is 400 metre long and its headwall is 65 metre high, while the middle cirque stairway is 500 metres long and 80 metres high. They are partially filled up by glacial debris. The upper cirque dips at an angle of 30° while the middle one dips at 18° towards the lower Shishnag cirque (Fig. 61). The second cirque stairway has been identified southwest of the lake at the altitudes of 3640 and 3725 metres respectively. The cirques dip towards Zamtirnag at 22° (Fig. 61A). The whole complex of Shishnag exhibit cirque-in-cirque form. The curved backwall has been cut primarily in Panjal trap, while the primary cirque floor has been scoured in limestone. In the northwest and southeast of the cirque floor, the limestone has been eroded and new cirque floor has been scoured deep in the underlying limestones.

There are 2 well developed cirque lakes (tarns) in the region, the largest being Shishnag. It is 1500 metres long, 360 metres broad and 400 metres high from the lip (Fig. 62). It is located at an altitude of 3575 metres. At its outlet the cirque-lip is 68 metres high. The right

Table 19: *Correlation between Cirque Variables of the Shishram Mountain Group*

Sl. No.	Station	Direction of cirque floor	Elevation of cirque lip (metres, (above sea level)	Length of cirque floor (metres)	Breadth of cirque floor (metres)	Height of the cirque from cirque lip (metres)	Ratio between length and height	Ratio between length and breadth
1.	Gratbal	295°	4100	250	115	190	1.3	2.2
2.	Gratbal	315°	4010	175	75	55	3.2	2.3
3.	Gratbal	305°	3920	220	70	60	3.7	3.1
4.	Gratbal	328°	3820	265	95	100	2.6	2.8
5.	Mahagunus	185°	4120	125	65	40	3.1	2.0
6.	Mahagunus	180°	3800	165	55	45	3.6	3.0
7.	Mahagunus	175°	3780	200	60	50	4.0	3.3
8.	Mahagunus	305°	3700	250	80	65	2.8	3.1
9.	Gulol Gali	345°	3900	95	35	30	3.1	2.7
10.	Gulol Gali	330°	3820	145	42	32	4.5	3.4
11.	Zamtimag	350°	3750	124	38	58	2.1	3.3
12.	Zamtimag (Cirque Lake)	325°	3720	110	26	35	3.1	4.2
13.	Zamtimag (Cirque Lake)	5°	3700	95	22	40	2.4	4.3
14.	Zamtimag Lake	330°	3640	920	312	212	4.3	2.9
15.	Shishnag Cirque Glacier	342°	4120	4500	2143	1552	2.9	2.1
16.	Cirque	30°	3680	445	200	90	5.0	2.5
17.	Cirque Lake	305°	3575	1500	400	290	5.2	3.7
18.	Saskat	295°	4000	200	80	70	3.0	2.5
19.	Saskat	300°	3940	95	40	45	2.1	2.4
20.	Saskat	320°	3680	325	100	80	4.0	3.3
21.	Kannag (Cirque Lake)	15°	3670	410	145	180	2.3	2.9
	MEAN		3832	504	200	158	3.2	3.0

Source: Field Observations.

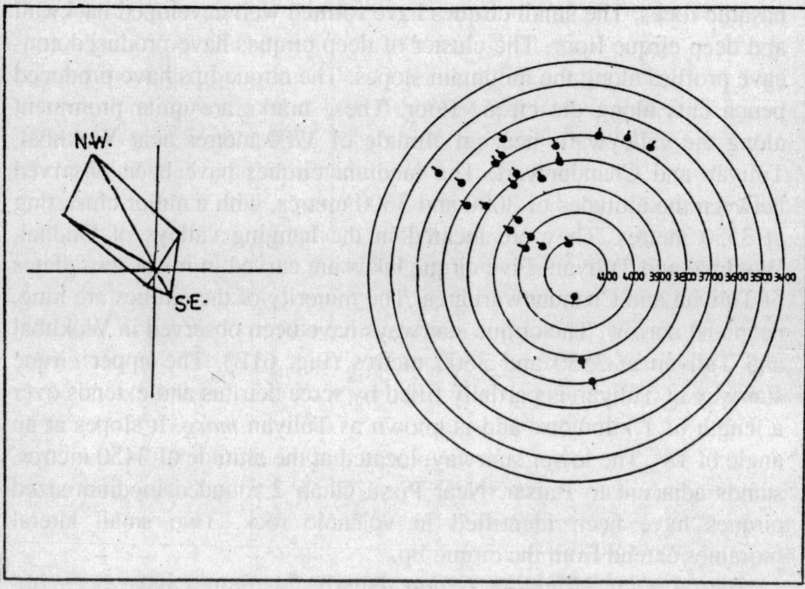

Fig. 60: *Shishram Mountain Group Orientation of Cirques*

bank of the cirque exposes the limestone rocks, while the left bank is covered with boulder and gravels with bedrocks of limestone at higher elevation. The backwall of the cirque is mainly composed of basaltic rocks. The second cirque lake is located adjacent to Shishnag at an altitude of 3650 metres. It has produced a subdued headwall and well developed side walls. The cirque-lip is breached in the centre by vertical notch.

East Liddar Mountain Group: Nearly 18 cirques have been identified in the East Liddar mountain group, of which 6 are small, 11 of medium size and 3 are large (Fig. 57C). Six cirques have produced lakes and 5 cirque glaciers have originated from them. The majority of the cirque glaciers are located above the altitude of 3850 metres. The icefree cirques are located between the altitude of 2200 and 3800 metres (Table 20). About 78 per cent of the total cirques located at the mean elevation of 3540 metres have north orientation (Figs. 58 and 63). The majority of the small cirques confined within a narrow altitudinal zone of 3650 metres and 3720 metres are mostly excavated in basaltic rocks. The small cirques have formed well developed backwall and deep cirque floor. The cluster of deep cirques have produced concave profiles along the mountain slopes. The cirque lips have produced bench cuts along the cirque floor. These marks are quite prominent along the valleywalls near an altitude of 3700 metres near Wokhbal, Tuliyan and Chandanwari. The medium cirques have been observed between the altitudes of 3000 and 3600 metres, with a major clustering at 3550 metres. They are located in the hanging valleys of Dudhal, Wokhbal and Tuliyan. Five cirque lakes are carved in micaceous slates of Tuliyan and Chandanwari area. The majority of the cirques are long, deep and narrow. The cirque stairways have been observed in Wokhbal and Tuliyan at 3530 and 3600 metres (Fig. 61B). The upper cirque stairway of Tuliyan is partially filled by scree detritus and extends over a length of 1 kilometre and is known as Tuliyan *marg*. It slopes at an angle of 18°. The lower stairway, located at the altitude of 3450 metres, stands adjacent to Baisar. Near Pissu Ghati 2 rounded medium-sized cirques have been identified in volcanic rock. Two small lateral moraines extend from the cirque lip.

West Liddar Mountain Group: Nearly 24 cirques have been observed in the West Liddar mountain group, and nearly half of them have developed into cirque lakes (Fig. 57D). There are neither cirque glaciers nor the small cirques, because most of the glaciers have retreated during the third interstade. There are 20 medium-sized and 4 large cirques in this mountain groups (Table 21). About 58 per cent of the cirques are located at a mean elevation at 3513 metres (Table 22) (Fig. 58). They have south-north orientation. About 42 per cent of the cirques located at 3350 metres have north-south orientation (Table 23) (Fig. 64). The cirques are located between altitudes of 2800 and 3700 metres. Two cirque clusters have been identified at altitudes of 3615 and 3250 metres, respectively (Fig. 63). At 3615 metres most of the large cirques on carved shale and basalt have produced bench cuts on valley sides. These marks are prominent near Satlanjan, Tarsar and Marsar sites. The other cluster of medium cirques have been identified at an altitude of 3250 metres. The cirques have been cut in quartzite, greywacke and basalt rocks. The cirque floors are produced on greywacke bed and lips in quartzite rock. The coalescence of cirque have joined the lips together and developed a bench cut on valley side.

Six cirques have been identified between Liddarwat and Basmai anticline, they have produced irregular floor but well developed lip.

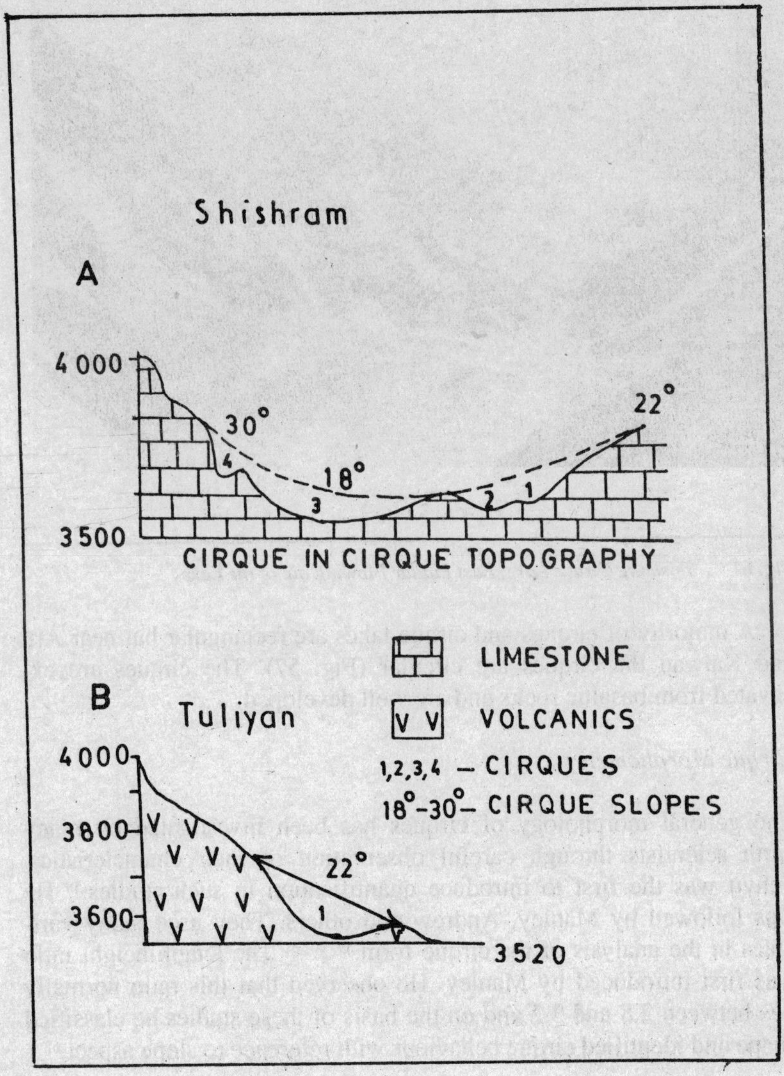

Fig. 61: *Cross-section of Cirque Lakes*

Fig. 62 *Shishnag Cirque Lake. East Liddar Flowing out of the Lake*

A majority of cirques and cirque lakes are rectangular but near Aru and Kalwan the cirques are circular (Fig. 57). The cirques are excavated from basaltic rocks and are well developed.

Cirque Morphometry

The general morphology of cirques has been investigated by many earth scientists through careful observation of their characteristics. Schytt was the first to introduce quantification in such studies.[58] He was followed by Manley, Andrew and others. They used many variables in the analysis of the cirque form.[59, 60, 61] The length-height ratio was first introduced by Manley. He observed that this ratio normally lies between 2.8 and 3.2 and on the basis of these studies he classified shape and identified cirque behaviour with reference to slope aspect.[62]

Table 20: *Correlation between Cirque Variables of the East Liddar Mountain Group*

Sl. No.	*Station*	*Direction of cirque floor*	*Elevation of cirque lip (metres, above sea level)*	*Length of cirque floor (metres)*	*Breadth of cirque floor (metres)*	*Height of the cirque from cirque lip (metres)*	*Ratio between length and height*	*Ratio between length and breadth*
1.	Kaltip	30°	4100	150	40	65	2.3	3.8
2.	Sonsar	128°	3920	950	240	500	2.0	4.0
3.	Sonsar	355°	3780	200	60	65	3.1	3.3
4.	Sonsar (Cirque Lake)	35°	3760	800	200	390	2.1	4.0
5.	Dudhal	335°	3850	1200	350	430	2.8	3.4
6.	Dudhal	325°	3600	225	70	50	4.5	3.2
7.	Chalwas	135°	3610	300	90	85	3.4	3.3
8.	Chalwas	320°	3000	185	55	60	3.1	3.4
9.	Hagan	185°	3810	235	58	86	2.9	4.0
10.	Wokhbal Cirque Glacier	300°	3900	985	133	352	2.8	7.4
11.	Wokhbal Cirque	31°	3700	170	50	62	2.8	3.4
12.	Sazakatyar	350°	3590	220	50	84	2.8	4.4
13.	Pahalgam	25°	2200	1500	478	245	6.1	3.1
14.	Tuliyan Cirque Glacier	348°	4000	950	185	535	1.6	5.1
15.	Tuliyan Cirque	160°	3600	425	100	150	2.9	4.3
16.	Tuliyan Cirque Lake	345°	3530	500	120	225	2.2	4.2
17.	Tuliyan	272°	3500	200	40	80	2.5	5.0
18.	Tuliyan	325°	3050	75	15	30	2.5	5.0
	MEAN		3583	515	129.7	194.1	2.9	4.1

Source: Field Observations.

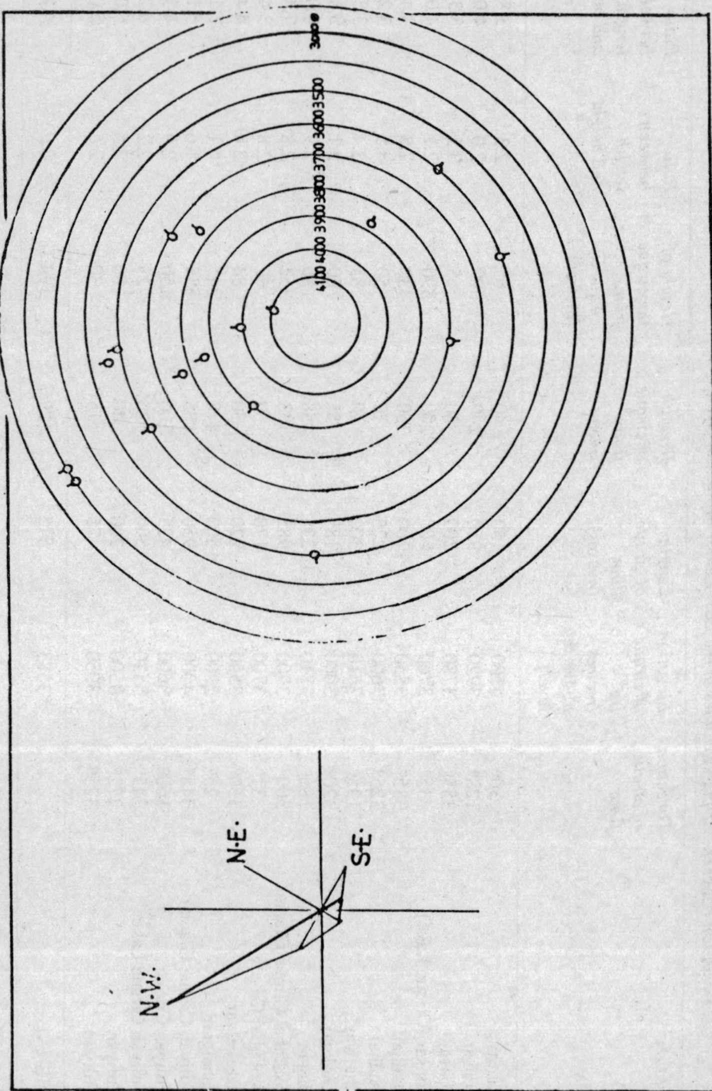

Fig. 63: *East Liddar Mountain Group Orientation of Cirques*

Table 21: *Correlation between Cirque Variables of the West Liddar Mountain Group.*

Sl. No.	*Station*	*Direction of cirque floor*	*Elevation of cirque lip (metres) above sea level)*	*Length of cirque floor (metres)*	*Breadth of cirque floor (metres)*	*Height of the cirque from cirque lip (metres)*	*Ratio between length and height*	*Ratio between length and breadth*
1.	Sanasar	185°	3700	300	80	100	3.0	3.8
2.	Cirque Lake	170°	3640	900	300	250	3.6	3.0
3.	Basmai	182°	3560	500	260	120	4.2	1.9
4.	Satlanjan	155°	3275	410	280	80	5.1	1.5
5.	Dodasar	140°	3600	300	80	100	3.0	3.8
6.	Dodasar Cirque Lake	182°	3640	375	250	180	2.1	1.5
7.	Handutsar	30°	3610	100	75	50	2.0	1.3
8.	Handutsar Cirque Lake	60°	3600	500	100	150	3.3	5.0
9.	Handutsar	65°	3580	850	320	220	3.8	2.6
10.	Chandansar	100°	3600	150	25	45	3.3	6.0
11.	Chandansar	105°	3400	400	130	300	1.3	3.1
12.	Chandansar Cirque Lake	108°	3200	500	180	150	3.3	2.8

(*Contd.*)

Table 21: *Contd.)*

Sl. No.	Station	Direction of cirque floor	Elevation of cirque lip (metres) above sea level)	Length of cirque floor (metres)	Breadth of cirque floor (metres)	Height of the cirque from cirque lip (metres)	Ratio between length and height	Ratio between length and breadth
13.	Tarsar	45°	3650	200	45	80	2.5	4.4
14.	Tarsar Cirque Lake	52°	3580	1850	650	300	6.2	2.8
15.	Chandbari	185°	3200	400	150	80	5.0	2.7
16.	Chumnai	12°	3650	300	100	100	3.0	3.0
17	Chumnai Cirque Lake	15°	3520	350	150	100	3.5	2.3
18.	Chumnai	20°	3500	150	75	40	3.8	2.0
19.	Nandain	35°	3100	250	100	150	1.6	2.5
20.	Tsonsar	40°	3500	400	140	110	3.6	2.9
21.	Tsonsar	45°	2700	300	180	140	2.1	1.7
22.	Ganshipathar	55°	2600	500	300	100	5.0	1.7
23.	Ganshipathar	35°	2400	250	100	80	3.1	2.5
24.	Munawarsar	20°	3200	400	160	180	2.2	2.5
	MEAN		3375	443	176.3	133.6	3.4	2.7

Source: Field Observations.

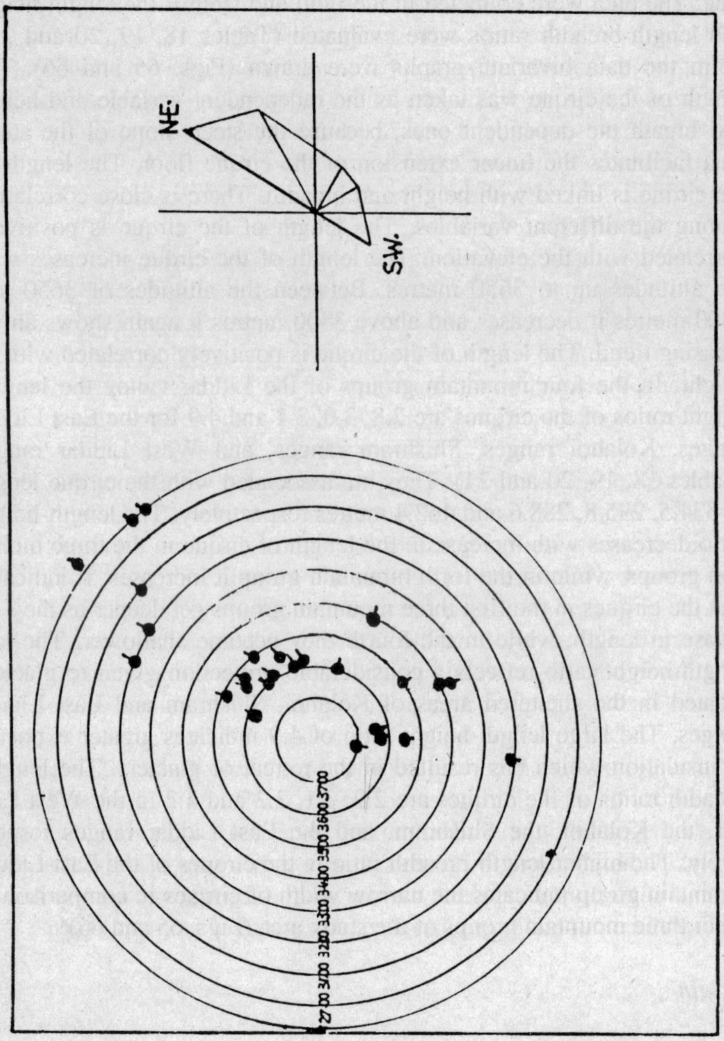

Fig. 64: *West Liddar Mountain Group Orientation of Cirques*

The author has used five variables : length, breadth, height, and the ratios of length/height and length/breadth to describe the shape and size of individual cirque, distributed in four mountain groups of the area. The data were collected in the field and from it the length-height and length-breadth ratios were evaluated (Tables 18, 19, 20 and 21). From the data bivariate graphs were drawn (Figs. 65 and 66). The length of the cirque was taken as the independent variable and height and breath the dependent ones, because the steep slope of the study area facilitates the linear extension of the cirque floor. The length of the cirque is linked with height and breadth. There is close correlation among the different variables. The length of the cirque is positively correlated with the elevation. The length of the cirque increases with the altitudes up to 3650 metres. Between the altitudes of 3650 and 3800 metres it decreases and above 3800 metres it again shows an increasing trend. The length of the cirque is positively correlated with its height. In the four mountain groups of the Liddar valley the length-height ratios of the cirques are 2.8, 3.0, 3.1 and 4.9 for the East Liddar ranges, Kolahoi ranges, Shishram ranges, and West Liddar ranges (Tables 18, 19, 20 and 21). They are associated with the cirque length of 334.5, 295.8, 288.6 and 462.4 metres respectively. The length-height ratio decreases with increase in the length of cirque in the three mountain groups, while in the forth mountain group it increases. It indicates that the cirques in the first three mountain groups get deeper as they increase in length, while in the fourth they become shallower. The low length-height ratio reflects a considerable protection given to glaciers located in the sheltered areas of Kolahoi, Shishram and East Liddar ranges. The large length-height ratio of 4.9 indicates greater exposure to insulation which has resulted in the retreat of glaciers. The length-breadth ratios of the cirques are 2.9, 3.1, 3.2 and 4.2 in the West Liddar, the Kolahoi, the Shishram, and the East Liddar ranges respectively. The higher length-breadth ratio in the cirques of the East Liddar mountain group indicates the narrow width of cirques in comparison to other three mountain groups of the study area (Figs. 65 and 66).

Origin

The majority of the small cirques are well developed. Their backwalls and lips have suffered only a little change. They are cut in volcanic massif of broad mountain summits lying in the northern part of Liddar valley. The morphogenesis of the cirques reveal their recent origin.

The medium and large cirques are somewhat irregular in shape and in some cases their headwalls and cirque lips are weathered (Fig. 53). They are much older in age than the small cirques. The striking difference in size between the large and medium cirque is indicated by the fact that the average length of the former is 1040 metres and that of the latter is 480 metres only. The greatest erosional potentials were available in the large cirques. The large cirques are located mostly near the valleyheads of the East and the West Liddar at an altitude of 3550 metres and medium cirques below 3400 metres. They clearly reflect the original location in the water eroded hollows of preglacial valleyhead and on the sides of preglacial trough.[63, 64, 65] Probably the hollows of the cirques were formed during the preglacial period by differential erosion. The fluted markings have been observed along the cirque walls of Shishnag, Tarsar, and Tuliyan lakes above the lake height of 200 to 250 metres. Grinlinton has also reported such markings all along the walls of the East Liddar valley. This further supports the view that the initiation of the process of hollow formation of medium and large cirques took place during preglacial time. Such hollows were cut at the contact zone of volcanics with arenaceous and limestone rocks. The majority of the large cirque lakes are located at an altitude of 3550 metres (Table 22). They are situated either near the confluent zone of glaciers of near the drifted snow sites. The unusually large Shishnag cirque (nearly 2 kilometres long) was excavated by the confluence of 8 cirque glaciers. Six glaciers came from northwest and two from northeast. They converged near the preglacial hollow at an altitude of 3800 metres. Similarly the Tarsar and Tuliyan lakes were formed due to drifting of snow from the adjacent cirque glaciers.

The majority of the cirques identified above the altitude of 2800 metres contain screes on their polished cirque floor. Small moraines extend below the cirque lip. The presence of scree and moraines suggests that the cirques have been formed by the combined action of mechanical weathering and glacial erosion. The backwalls of the cirques have been formed by alternate processes of freezing and frost-shattering.[66, 67, 68] The higher slopes of the Liddar ranges are steep (60°-70°) and could not retain glacial ice. Therefore, glacial ice slipped into preglacial hollows through the basal slippage from their valleyhead. The slippage led to the formation of basal meltwater. The meltwater percolated through the joints of volcanics, sandstone, slates, and calcareous rocks. The joints are spaced between 10 and 100 centimetres.

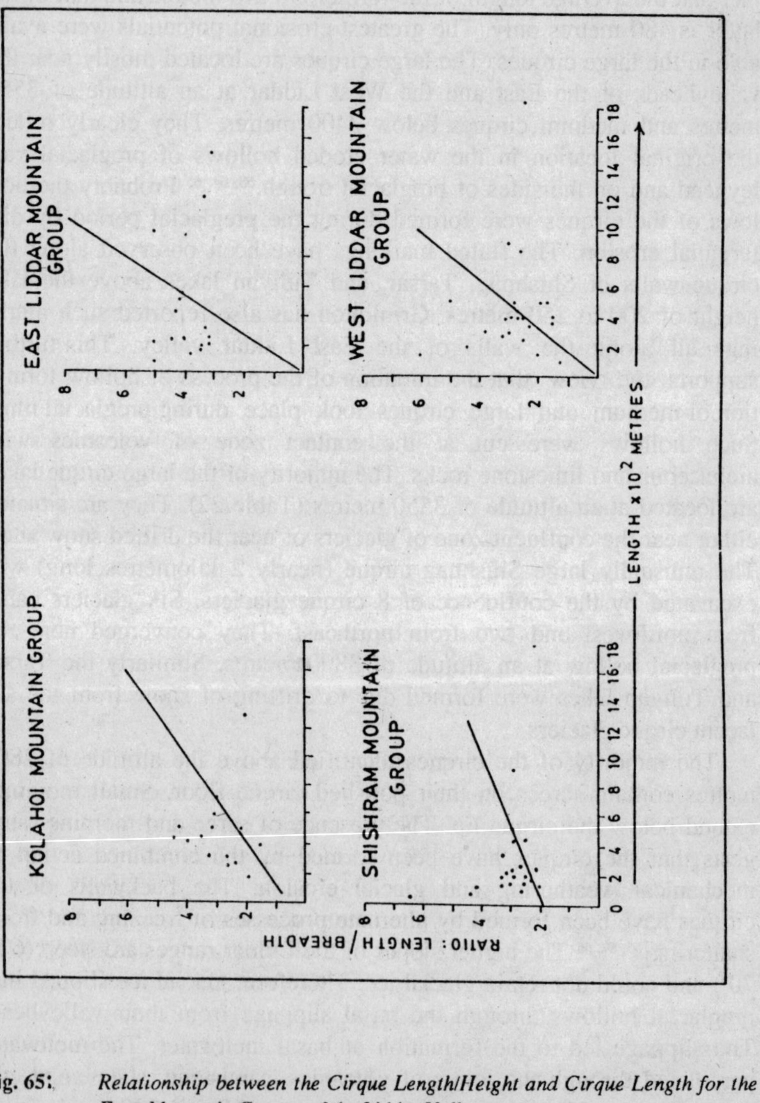

Fig. 65: *Relationship between the Cirque Length/Height and Cirque Length for the Four Mountain Groups of the Liddar Valley*

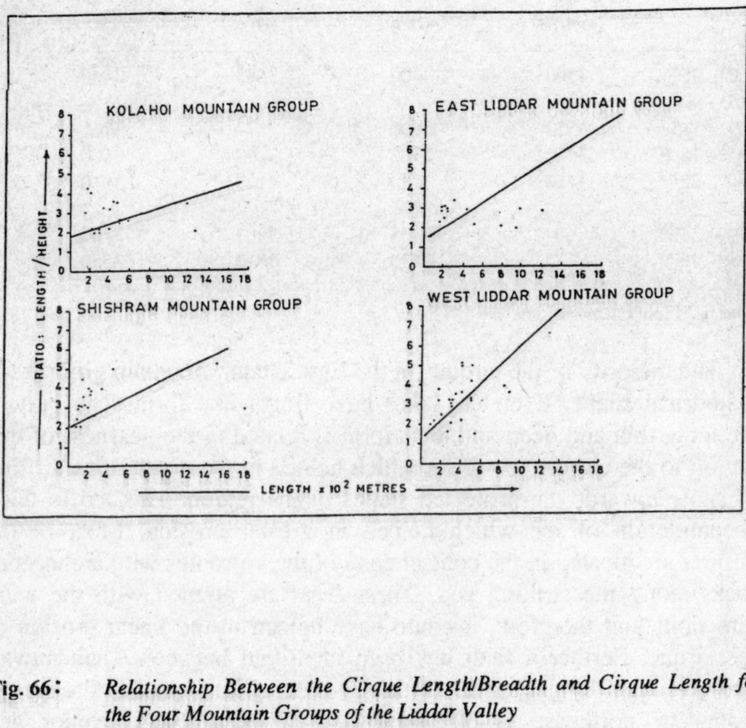

Fig. 66: *Relationship Between the Cirque Length/Breadth and Cirque Length for the Four Mountain Groups of the Liddar Valley*

The parts of the headwalls formed from the jointed rocks were exposed to low temperature below freezing point. The water in the joints of the rocks caused 9 to 12 per cent dilation during freezing. It ultimately helped frost shattering.[69] The dilated joints at the base of the headwall were attacked by glacial ice and rock fragments were plucked from it. The scouring was facilitated by rotational movement of glacier near the base of the headwall. The rotation movement led to vertical deepening and linear extension of cirque floors.[70, 71, 72] The length, breadth and height ratios have given different geometrical shape to the cirques. In the study region the shape of cirque varies from cylindrical to rectangular.

Table 22: *Cirque Lip Elevation and Frequency by Orientation Class in the Four Mountain Groups of the Liddar Valley*

Orientation	*West Liddar Mountain Group (metres)*	*Kolahoi Mountain Group (metres)*	*Shishram Mountain Group (metres)*	*East Liddar Mountain Group (metres)*
0° - 45°	3500	3400	3683	3440
45° - 90°	3525	—	—	—
90° - 135°	3533	3600	—	3765
135° - 180°	3050	3700	3790	3600
180° - 225°	3450	3720	4120	3810
225° - 270°	—	3900	—	—
270° - 315°	—	3405	3893	3700
315° - 360°	—	3720	3810	3550

Source: Field Observations.

The majority of the cirques in the East Liddar mountain group have cylindrical shape. Even the lakes have finger-like form. The cirques are long, thin and deep and their form is related to the nearness of this region to the eastern ice fence which helped in the continuous drifting of snow towards the protected slopes of the region. This led to thick accumulation of ice which helped in greater erosion. Most of the cirques are located in the contact zone of the volcanics with arenaceous rocks along the striking bed. These beds are aligned with the wind directions and therefore, it could have helped in the linear erosion of the cirque. Further a fault has been identified between Chandanwari and Pahalgam by geologists of the Geological Survey of India. It runs from 60° northeast to 60° southwest between the volcanics and arenaceous rocks. There has been a displacement of beds for nearly 1500 metres. This displacement could have resulted in greater erosion along the major axis of cirques than along its two sides.

The rectangular shape of the majority of the cirques in the West Liddar mountain ranges could be due to thin glacial ice cover and unfavourable wind direction. In this group only 58.3 per cent of the cirque have north orientation in comparison to the 76 per cent in the other three mountain groups.

In the Kolahoi and the Shishram mountain groups the cirques are developed on the steep slopes (40° to 60°) at higher altitude. The steep slopes have been instrumental in the formation of triangular cirques.

The Liddar cirque floors have dominant north and northwest orientation. Nearly 75 per cent of the total cirques, located in three mountain groups, have north orientation (Table 24) (Figs. 59, 60, 63 and 64). Only 58 per cent of the West Liddar mountain group cirque floors have north orientation. When we consider all the mountain groups we find that 56 per cent of the cirque floors have southeast-northwest orientation. The orientations reveal that the cirques were formed mainly on north and northwest slopes of the mountains where the sites favoured the snow accumulation. The snow accumulated largely through drifting and was redistributed by wind action in the sheltered areas. In the Pleistocene epoch the northwest slopes must have provided sheltered sites. It would appear that the dominant wind direction during the formation of cirque was probably northwest. The snow which fell on the southeast-facing slope would have been subjected to redistribution by wind due to the exposed site. The only wind direction data available for the nearest area comes from Srinagar and Gulmarg. This data can be assumed as almost similar to that of the study area. The dominant wind direction of the Srinagar and Gulmarg station show that its southeast-northwest orientation is similar to those of the cirques (Table 24).

Table 23: *Average Percentage of Cirques with Specific Orientation in the Four Mountain Groups of Liddar Valley*

Orientation	*West Liddar*	*Kolahoi*	*Shishram*	*East Liddar*
0° - 45°	37.5	22.4	14.5	22.4
45° - 90°	20.8	—	—	—
90° - 135°	12.5	5.5	—	11.2
135° - 180°	12.5	5.5	9.6	5.5
180° - 225°	16.7	11.2	4.8	5.5
225° - 270°	—	5.5	—	—
270° - 315°	—	11.2	33.1	11.2
315° - 360°	—	38.7	38.0	44.2

Source: Field Measurements.

Table 24A: *Frequencies of Wind Direction in Srinagar, 1891-1940*

Month	*Direction*								
	North	*Northeast*	*East*	*Southeast*	*South*	*Southwest*	*West*	*Northwest*	*Calm*
January	10.0	4.0	1.5	16.5	8.5	4.5	11.0	16.5	27.5
February	6.5	4.5	2.0	17.0	8.5	7.0	15.0	20.5	19.0
March	9.0	3.0	2.0	14.0	7.5	5.0	16.0	25.5	18.0
April	11.5	4.0	2.0	18.0	10.0	5.5	16.0	19.5	13.5
May	8.5	7.5	3.5	16.0	10.5	6.5	7.5	12.5	27.5
June	10.0	10.5	2.5	13.5	10.0	4.5	8.0	13.0	28.0
July	10.5	5.5	3.5	14.0	10.0	5.5	10.0	15.5	25.5
August	6.5	5.0	2.0	17.5	11.0	4.5	13.5	13.0	27.0
September	7.0	3.0	2.5	14.0	7.5	4.0	12.0	12.0	38.0
October	6.0	4.5	1.5	13.5	9.0	4.0	6.5	16.0	39.0
November	10.5	4.0	3.5	12.5	4.5	3.5	10.0	10.5	41.0
December	12.0	5.0	3.0	11.0	8.5	5.0	9.5	15.5	30.0
MEAN	9.0	5.5	2.5	15.0	9.0	4.5	11.0	15.5	28.0

Source: Climatological Tables of Observatories in India, 1952, p. 162.

Table 24B: *Frenquencies of Wind Direction in Gulmarg, 1901–1940*

Month	*Direction*								
	North	*Northeast*	*East*	*Southeast*	*South*	*Southwest*	*West*	*Northwest*	*Calm*
June	12	30	11	4	1	4	3	9	26
July	8	31	12	6	2	3	3	6	30
August	5	22	9	4	0	5	4	5	45
September	2	14	8	5	1	5	3	4	58

Source: Climatological Tables of Observatories in India, 1952, p. 164.

NOTES

1. King, C.A.M. (1980). 'Threshold in Glacial Geomorphology', *Thresholds in Geomorphology*, Donald, R. Coates and John, D. (Eds.), London; George Allen and Unwin Ltd., p. 297.
2. Grinlinton, J.L. (1928). 'The Former Glaciation of the East Liddar Valley, Kashmir', *Memoir, Geological Survey of India*, Vol. 49, part 2, pp. 289-388.
3. Ahmad, N. and Hashimi, N.H. (1974), 'Glacial history of Kolahoi Glacier, Kashmir, India', *Journal of Glaciology*, Vol. 13, pp. 279-283.
4. See ref. 2, p. 295.
5. Vivian, R. (1970). *'Hydrologie et erosion sousglaciaires'*, *Revue. Geog. alp.*, Vol. 58, Part 2, pp. 241-64.
6. Boulton, G.S. (1974). 'Process and patterns of glacial erosion', *Glacial geology*, Coates, D.R. (Ed.), Binghampton; State University of New York, pp. 41-87.
7. Embleton, C. (1980). 'Glacial processes', *Processes in Geomorphology*, Embleton, C. and Thornes, J. (Eds.), India : Arnold Heinemannde, pp. 272-306, ref. p. 279.
8. See ref. 6, pp. 64-66.
9. Rothlisberger, H. (1968). 'Erosive processes which are likely to accentuate or reduce the bottom relief of valley glaciers', *International Association of Scientific Hydrology*, Vol. 79, pp. 87-97.
10. Boulton, G.S. and Vivian, R. (1973). 'Underneath the glaciers', *Geographical Magazine*, Vol. 45, Part 4, pp. 311-319.
11. Andrews, J.T. (1975). *Glacier Systems : An approach to glaciers and their environments*, California; Wadsworth Publishing Company, ref. p. 118.
12. Matthes, F.E. (1930). 'Geological history of the Yosemite valley'. *U.S. Geological Survey Professional paper*, No. 160, pp. 137.
13. Crosby, I.B. (1945). 'Glacial erosion and the burried Wyoming valley of Pennsylvania', *Bulletin Geological Society of America*, Vol. 56, pp. 389-400.
14. Zumberge J.H. (1955). 'Glacial erosion in tilted rock layers', *Journal of Geology*, Vol. 63, pp. 149-158.
15. Trainer, F.W. (1973). 'Formation of joints in bedrock by moving glacial ice', *U.S. Geological Survey Journal Reasearch*, Vol. 1, Part 2, pp. 229-236.
16. Lewis, W.V. (1954). 'Pressure release and glacial erosion', *Journal of Glaciology*, Vol. 2, No. 16, pp. 417-422.
17. See ref. 6, p. 46.
18. Johnson, W.D. (1904). 'Profile of maturity in Alpine glacial erosion', *Journal of Geology*, Vol. 12, pp. 569-578, ref. p. 572.
19. Tricart, J. and Cailleux, A. (1962). *Le modele glaciaire et nival*, Paris; SEDS, p. 508.
20 Bakker, J.P. (1965). 'A forgotten factor in the interpretation of glacial stairways', *Zeitschrift fur Geomorphologie*, Neuefolge 9, pp. 18-34, ref. pp. 21-22.
21. Feininger, T. (1971). 'Chemical weathering and glacial erosion of crystalline rocks and the origin of till', *U.S. Geological Survey professional paper*, No. 750-C, pp. 65-81.
22. Thomas, M.F. (1974). *Tropical Geomorphology*, London; Macmillan, pp. 1-332.

23. Ollier, C.D. (1969). *Weathering*, Edinburgh; Oliver and Boyd, pp. 1-304, ref. pp. 28-30.
24. Galibert, M.G. (1962). 'Recherches sur les processes d'erosion glaciers de la Haute Montagne Alpine', *Bulletin Association, Geographer Francais*, Vol. 303, No. 4, pp. 8-46.
25. Chapman, C.A. and Roux, R.L. (1958). 'Statistical study of topography, sheeting and jointing in granite : Acadia National Park, Maine', *American Journal of Sciences*, Vol. 256, pp. 111-127.
26. Sugden, D.E. (1968). 'The selectivity of glacial erosion in the Cairngorm mountains. Scotland', *Transactions and Papers, Institute of British Geographers*, No. 45, pp. 79-80.
27. See ref. 9, pp. 87-90.
28. Boulton, G.S. (1972a). 'The role of thermal regime in glacial sedimentation'. *Institute of British Geographers special publication*, No. 4, pp. 1-19.
29. Chamberlin, T.C. (1988). 'The rock scourings of great ice invasions', *U.S. Geological Survey 7th Annual Report*, pp. 147-248.
30. Peterson, D.N. (1970). 'Glaciological studies on the Casement glacier, southeast Alaska', *Institute of Polar Studies*, Report No. 36. pp. 1-161.
31. See ref. 7, p. 199.
32. Charlesworth, J.K. (1957). *The Quaternary Era*, Vol. 1, London; Edward Arnold, pp. 1-900.
33. Linton, D.L. (1963). 'The forms of glacial erosion', *Transactions and Papers, Institute of British Geographers*, No. 33, pp. 1-28.
34. See ref. 12, pp. 125-128.
35. Flint, R.F. (1971). *Glacial and Quaternary Geology*, New York; John Wiley, pp. 1-892, ref. p. 568.
36. Clayton, K.M. (1965). 'Glacial erosion in the Finger Lakes region (New York State, U.S.A.)', *Zeitschrift fur Geomorphologie*, Neie Folge September, Band 9, pp. 50-62.
37. See ref. 2, p. 305.
38. Clark, J.M. and Lewis, W.V. (1951). 'Rotational movement in cirque and valley glaciers', *Journal of Geology*, Vol. 59, p. 6.
39. Lewis, W.V. (1947b). 'Valley steps and glacial valley erosion', *Transactions and Papers, Institute of British Geographers*, No. 17, pp. 19-44.
40. King, C.A.M. (1959). 'Geomorphology in Austerdalen, Norway', *Geographical Journal*, Vol. 125, pp. 357-369.
41. Rudberg, S. (1973). 'Glacial erosion forms of medium size — a discussion based on four Swedish case studies', *Zeitschrift fur Geomorphologie*, Band 17, pp. 33-48.
42. Svensson, H. (1959). 'Is the Cross-section of glacial valley a parabola?', *Journal of Glaciology*, Vol. 3, No. 25, pp. 362-363.
43. Hack, J.T. (1973a). 'Drainage adjustment in Appalachians' *Fluvial Geomorphology*, Morisawa, M. (ed.), New York; State University of New York, Binghamton, pp. 51-69.
44. Seebar, L. and Gornitz, V. (1983). 'River profiles along the Himalayan arc as indicators of active tectonics', *Tectonophysics*, Vol. 92, pp. 335-367.
45. See ref. 43, pp. 54-56.
46. See ref. 44, pp. 344-346.

47. Graf, W.L. (1970). 'The geomorphology of the glacial cross-section', *Arctic and Alpine Research*, Vol. 2, pp. 303-312.
48. Cotton, C.A. (1941). 'The longitudinal profiles of glaciated valleys', *Journal of Geology*, Vol. 49, pp. 113-128, ref. pp. 122.
49. Seddon, B. (1954). 'Late-glacial cwm glaciers in Wales', *Journal of Glaciology*, Vol. 3, No. 22, pp. 94-99.
50. Sugden, D. (1969). 'The age and forms of corries in the Cairngorm mountains, Scotland', *Scotish Geographical Magazine*, Vol. 83, pp. 34-46.
51. Andrews, J.T. (1965). 'The Corries of the Northern Nain-Okak section of Labrador', *Geographical Bulletin*, Vol. 7, pp. 129-136.
52. Trenhaile, A.S. (1975). 'Cirque elevations in the Canadian Cordillera', *Annals of Association of American Geographers*, Vol. 65, No. 4, pp. 518-529.
53. McCabe, L.H. (1939). 'Nivation and corries erosion in West Spitsbergen', *Geographical Journal*, Vol. 94, pp. 447-465.
54. Krenek, L. and Bhawan, U., (1945). 'Recent and past glaciation of Lahaul', *Indian Geographical Journal*, Vol. 20, Part 3, pp. 93-102.
55. Vohra, C.P. (1981). 'Himalayan Glaciers', *The Himalaya : Aspect of Change*, Lall, J.S. and Moddie, A.D. (Eds.), Delhi; Oxford University Press, pp. 138-151, ref. p. 144.
56. Ahmad, N. and Mayewski, P.A. (1981). 'Himalayan glaciation : An unresolved issue', *The Geographer*, Vol. 28, No. 2, pp. 21-31, ref. pp. 27-28.
57. Andrew, J.T. and Dugdale, R.E. (1971). 'Quaternary history of northern Comberland Penninsula, Baffin Island, N.W.T. Part V: factors affecting corrie of glacierisation in Okoa Bay', *Quaternary Research*, Vol. 1, pp. 532-551.
58. Schytt, V. (1959). 'The glaciers of the Kebnekajse missif', *Geografiska Annaler*, Vol. 41, pp. 213-217.
59. See ref. 11, pp. 137-142.
60. Haynes, V.M. (1968). 'The influence of glacial erosion and rock structure on corries in Scotland', *Geografiska Annaler*, Vol. 50A, pp. 221-234, ref. pp. 229-231.
61. Manley, G. (1959). 'Interpreting the meteorology of the late and post-glacial', *Palaeogeogr. Palaeoclimatol. Palaeoecol*, Vol. 10, pp. 163-175.
62. See ref. 19, pp. 512-513.
63. Davis, W.M. (1900). 'Glacial erosion in France, Switzerland and Norway', *Proceedings of Boston Society of Natural History*, No. 29, pp. 273-321.
64. Penck, A. (1905). 'Glacial features in the surface of the Alps', *Journal of Geology*, Vol. 13, No. 1, pp. 1-17.
65. Lewis, W.V. (1960). 'Norweigian cirque glaciers', *Royal Geographical Society Research Series*, No. 4, pp. 1-104.
66. See ref. 18, pp. 572-576.
67. Holmes, C.D. (1944). 'Hypothesis of glacial erosion', *Journal of Geology*, Vol. 52, pp. 184-190.
68. See ref. 39, p. 42.
69. See ref. 60, p. 234.
70. See ref. 38, pp. 554 - 560.
71. Tiku, A.K., Krishnamurthy, A.C. and Gupta, U.P. (1974). 'Report on Liddar Valley Hydel Scheme', *Geological Survey of India*, No. 35, pp. 1-30.
72. Embleton, C. and King, C.A.M. (1975). *Glacial Geomorphology*, London; Edward Arnold, pp. 181-205.

5

Geomorphological Features of Glacial Transportation

The movement of the Liddar glaciers, the East and the West Liddar, from the valleyheads to Pahalgam-Ganshibal basin followed different routes in response to gravity, differential solar radiation and fluctuations in mass balance (Fig. 67). In their movement the glaciers acquired subglacial, supraglacial and preglacial debris from Valleyheads. The subglacial material was produced by the erosion of the valleyfloor and valleywalls by the glaciers. The supraglacial debris was received from the material that falls on the ice surface from the adjacent nunataks, valleysides, cirque headwalls and talus slopes. The Liddar glaciers also transported the preglacial material inherited from the preglacial landscape. The glaciers transported a large volume of large rock fragments (erratics) and boulders in the downvalley direction. There was little resistance from the debris due to steep slope and little frictional drag. The more important of the geomorphic features produced by glacial transportation are erratics and boulders and pebbles.

ERRATICS AND BOULDERS

A large percentage of rock fragments and large boulders (with diameter of 3 metres) are present in the moraines of Aru, Chandanwari and Pahalgam. These huge rock fragments could not have been plucked by glaciers alone, but must have been produced from weathered supraglacial rock faces (Figs. 68 and 69). This is further supported by the identification of large erratic fields and perched boulders at Shishnag-Chandanwari and Aru-Liddarwat section of the East and West Liddar valley respectively (Fig. 68). Some of the perched boulders observed near the localities of Liddarwat, Mandlun, Chandanwari and Pahalgam are more than 10 metres long.

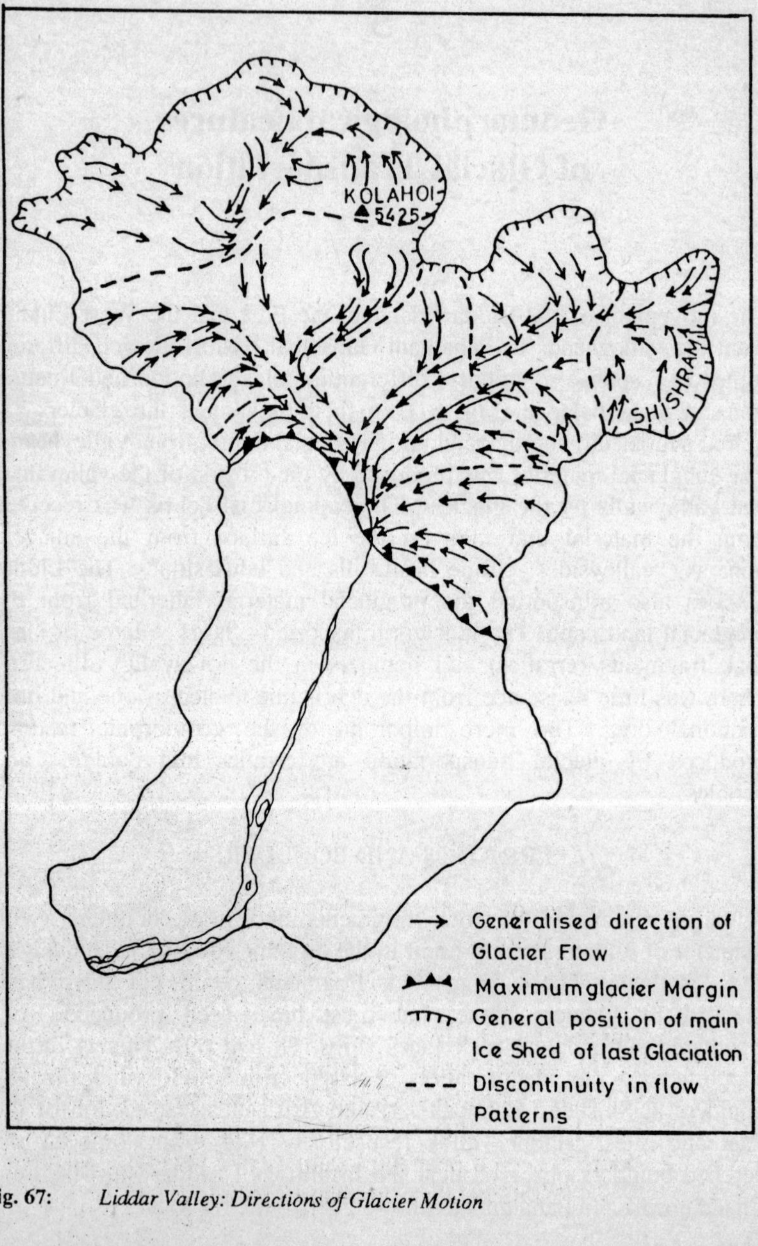

Fig. 67: *Liddar Valley: Directions of Glacier Motion*

SOURCE AREA OF THE BOULDERS AND ERRATICS

Most of the boulders and erratics comprise of volcanics derived from beds located at higher altitudes and exposed all along the valley sides. The agglomeratic slates and calcareous sandstone which form the bulk of the series are little represented in the erratics and perched boulders. The fragments readily disintegrate to form argillaceous matrix of boulder-clay, which is prevalent in tills of the moraine. The exposed volcanics at higher altitudes are racked in folds, so that the outcrops are repeated many times and in consequence cannot, at present, be used to restrict the source of any particular valley or locality. Their location at higher altitude indicates their origin in particular environment. The perched boulders, erratics and large boulders of the moraine constitute nearly 60 per cent of the transported material. Their possible source is from supraglacial debris.

Fig. 68A: *Large angular rock fragments near Kolahoi Ganj locality. The large size of volcanics and quartzitic rock fragments show their origin to supraglacial means*

Fig. 68B: *Typical isolated volcanic perched boulders (A) near Liddarwat. The boulders are striated and polished*

Origin

Most of the erratics, perched boulders and large boulders of moraines are rounded and polished (Fig. 71). They have been formed by englacial processes. Their large size and high sphericity reflect the immense power of the transportation of the Liddar glaciers.[1] The Kolahoi and Shishram group of glaciers presently are located near the exposed valleyheads of the East and the West Liddar. They are nearly 35 kilometres away from the erratics and perched boulders. The glaciers are highly crevassed. During the third glacial advance and retreat the accumulation section of the glaciers could have acquired large rock fragments from the exposed weathered rock valleysides as rockfalls (Fig. 70). These fragments were buried deep in the sole of glacier and were carried englacially. The supraglacial materials fall continuously in the numerous deep crevasses that characterise the ablation zone and are transported englacially. Therefore, very little material was carried on the surface of the ablation zone.

SEDIMENT TRANSPORT DIRECTION

Much of the transported material deposited by glaciers show preferred orientation of the fabrics of the sediments. The long axes of the embedded pebbles lie parallel to the direction of ice flow.[2] Therefore, undisturbed tills have got inherent form of the transported sediments. The analysis of the pebble orientation of tills has developed as one of the common tools in evaluating the palaeogeographic environment for sediment transport direction.[3,4] Such studies can also be deduced by studying the orientation of striae and recessional moraines.[5,6,7,8]

Fig. 69A: *The erratic field near the summit of Shishnag mountain. The erratics are supposedly of Late Ganshibal stade*

Fig. 69B: *Boulder conglomerates of limestones near Zaijpal. The boulders are striated, rounded and have a preferred orientation to northwest*

PEBBLE ORIENTATION

The pebble orientation of the till deposits of the moraine ridges of the East and the West Liddar valley was measured during 1983-85 field seasons. Fifteen sample sites were chosen along the road cuts of the exposed sequence of moraine ridges from Ganshibal to Shishnag and Pahalgam to Kolahoi in the East and the West Liddar valley (Fig. 71). The measurements at each site were made in one square metre plots. All the long rod-like pebbles with axial ratio of 1:2 identified from each plot were removed from its position in the soil matrix. The orientation (dip) of the long axis of the pebble was identified from its impression and measured by Brunton compass. The pebbles touching one another or having vertical lineament were excluded. The remaining pebbles were found to have axial length of up to 15 centimetres but for most of them it ranged between 1 to 10 centimetres. From each of the 15 selected sites 100 pebbles were randomly picked up and their pebble directions were measured. The vectors with standard deviation of about 1° were plotted as function of orientation. The data from all the stations were plotted by composite rose diagram (Fig. 72). The standard deviation of the orientation of pebbles is very large (16° to 40°). At each site the vector mean was calculated from the standard deviation (Table 25).

Harrison test was applied in order to test whether the randomly chosen pebbles have yielded the preferred orientation. Harrison suggested that any randomly chosen three increments of 25 consecutive measurements from each site was sufficient to define the direction of the preferred orientation.[9, 10] Of the 15 sites from which 100 pebbles were picked up and their directions were measured, the five sites were randomly selected (Figs. 72B, I, D, M, and E). From each of these five sites the direction of 100 pebbles were already measured in the field. Out of these 100 measured pebbles the two increments of 50 and 25 pebbles were randomly selected and their vector mean were recalculated (Table 26).

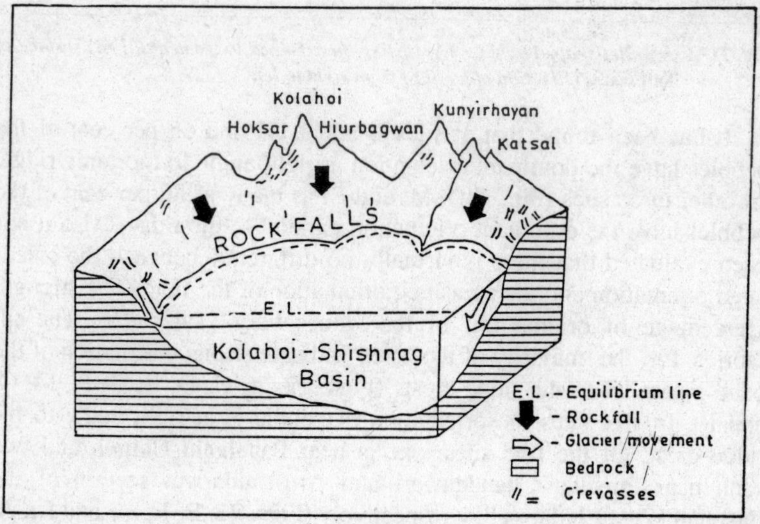

Fig. 70: *Rockfall Sites of Kolahoi and Shishnag Basins*

Fig. 71A: *A Bullet-shaped boulder (A) resting on exhumed lodgement till (B) surface of Satlanjan, Direction of ice-flow from right to left*

It has been found that at sites B and I, 55 and 65 per cent of the pebbles have the dominant orientation at right angle to morainic ridge. At other three sites (Fig. 72D, M, and E) as many as 80 per cent of the pebbles have the dominant orientation parallel to the ridge. Also it has been evaluated that there is virtually no difference between the calculated orientation and the measured orientation of the field. The strength (percentage of orientation) of the values vary (Table 26). The till fabrics for the majority of the sites indicate strong orientation of the pebble parallel to the ridge crests (Figs. 72, A, C, D, E, H, K, M, O, and P). In five sites the orientation of pebble is at right angle to the ridge crest. Of the five sites, one is near Pahalgam Church and two each near Parsulan-Chandanwari and Aru-Liddarwat section of the East and West Liddar valley respectively (Figs. 72, B, F, I, J and L). At these sites the dominant orientation of the pebbles changes as the orientation of the ridge crest changes. On the whole, the pebble orientation is northeast to southwest and northwest to southeast in the East and the West Liddar respectively (Table 26).

Origin of the Pebble Orientation

From the above studies it is evident that, by and large, fabrics in the tills were dumped by the ice as they show distinct parallelism with ice movement. The samples located near the localities of Pahalgam, Aru, Chandanwari and Liddarwat have pebbles at right angle to the ridge crest. The fabric diagrams indicate that a majority of the pebbles dip towards the proximal side of the ridge. The possible mechanism of formation could be due to the confluence of the main valley glacier with the hanging glacier resulting in upward movement of hanging ice along the thrust plane. This could have squeezed water-saturated till beneath the ice to form ridge, the crest line of which could be parallel to the ice front. Thus two dominant modes were responsible for their origin.

Fig. 71B: *Exposed sequence of Pharsulan lateral moraine shows polished boulders and pebbles. The orientation of boulders and pebbles is at right angle to the morainic ridge shown by arrows*

Fig. 71C: *The exposed sequence of Pahalgam lateral moraine. The till matrix is dominated by pebbles orientated parallel to moraine ridge. The impression of long axis of pebbles and striation line on rock fragments shown by arrows are aligned parallel*

Table 25: *Preferred Orientation of Pebbles in Liddar Till Matrix*

Site	*Number of pebbles*	*Standard deviation*	*Tabulated vector mean of pebbles*
A	100	19°	17°
B	100	20°	102°
C	100	22°	60°
D	100	30°	62°

(*Contd.*)

Table (*Contd.*)

E	100	24°	322°
F	100	24°	346°
G	100	30°	40°
H	100	30°	350°
I	100	20°	10°
J	100	27°	101°
K	100	16°	0°
L	100	17°	83°
M	100	31°	29°
O	100	40°	2°
P	100	30°	90°

Source: Field Observations.

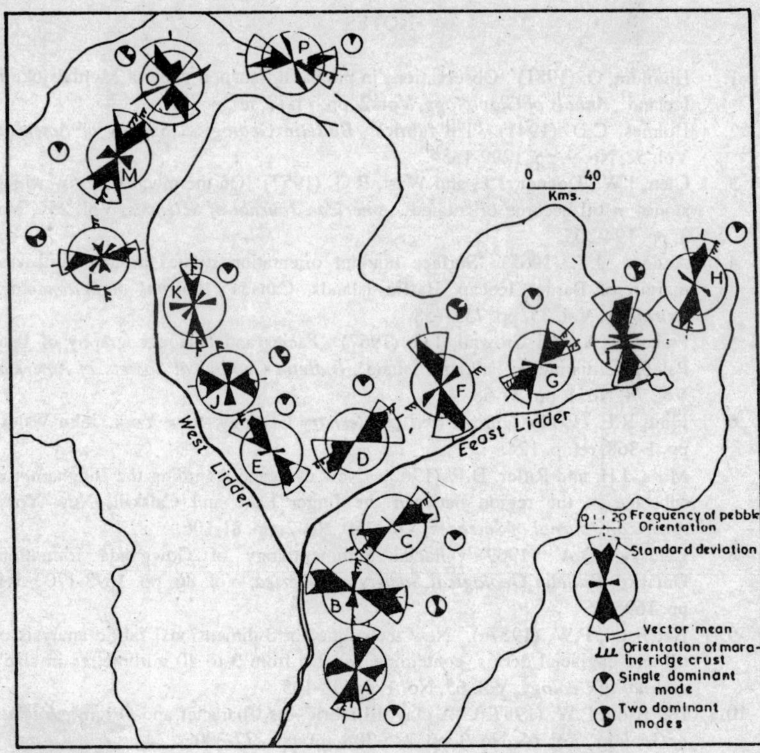

Fig. 72: *Liddar Valley Pebble Orientation in Moraines*

Table 26: *Comparative Pebble Numbers and their Calculated Orientation*

Site	100 Pebbles		50 Pebbles		25 Pebbles	
	Orientation	*Strength (per cent)*	*Orientation*	*Strength (per cent)*	*Orientation*	*Strength (per cent)*
B	E 12° S - W 12° N	65	E 10° S - W 10° N	60	E 11° S - W 11° N	58
I	N 80° W - S 80° E	59	N 85° W - S 85° E	55	N 88° W - S 88° E	64
D	N 62° E - S 62° W	77	N 61° E - S 61° W	85	N 60° E - S 60° W	80
M	N 29° E - S 29° W	62	N 32° E - S 32° W	78	N 33° E - S 33° W	81
E	N 52° W - S 52° E	76	N 58° W - S 58° E	83	N 60° W - S 60° E	74

Source: Field Observations.

NOTES

1. Humlum, O. (1981). 'Observations in the basal transport zone of Myrdalsjokull, Iceland', *Annals of Glaciology*, Vol. 2, pp. 71-77, ref. p. 74.
2. Holmes, C.D. (1941). 'Till fabrics', *Bulletin Geological Society of America*, Vol. 52, No. 9, pp. 1299-1354.
3. Glen, J.W., Donner, J.J., and West, R.G. (1957). 'On the mechanism by which stones in till become orientated', *American Journal of Sciences*, Vol. 255, No. 3, pp. 194-205.
4. Andrew, J.T. (1965). 'Surface boulder orientation studies around northwest margin of Barnes Icecap, Baffin Islands, Canada', *Journal of Sedimentary Petrology*, Vol. 35, pp. 735-758.
5. Frakes, L.A. and Crowell, J.C. (1967). 'Facies and Palaeogeography of Late Paleozoic diamicite, Falkland Island', *Bulletin Geological Society of America*, Vol. 78, No. 1, pp. 37-58.
6. Flint, R.F. (1971). *Glacial and Quaternary Geology*, New York, John Wiley, pp. 1-368, ref. p. 124.
7. Moss, J.H. and Ritler, D.F. (1962). 'New evidence regarding the Binghampton substage in the region between the finger Lake and Catskill, New York, *American Journal of Sciences*, Vol. 260, No. 2, pp. 81-106.
8. Lindsey, D.A. (1969). Glacial Sedimentology of Gowganda formation, Ontario', *Bulletin Geological Society of America*, Vol. 86, pp. 1693-1700, ref. pp. 1695-96.
9. Har rison, P.W. (1957a). 'New technique for 3-dimensional fabric analysis of till and englacial debris, containing particle from 3 to 40 millimetres in size', *Journal of Geology*, Vol. 65, No. 1, pp. 98-105.
10. Har rison, P.W. (1957b). 'A clay till fabric—Its Character and Origin', *Journal of Geology*, Vol. 65, No. 3, pp. 275-308, ref. pp. 277-300.

6

Geomorphological Features of Glacial Deposition

The glacial depositional features are created directly by the advance and retreat of glaciers. They, therefore, provide much of the information regarding the history of the Pleistocene glaciers. The advance produces distinctive depositional features which epitomise the changing times of landform formation. The retreat of the glaciers produces distinctive interwoven moraine features linked with the substages of glacier retreat. The volume of deposition is closely correlated with the intensity of weathering and erosion. The debris produced by these processes were carried to the glacier sole, followed by transport of material by subglacial and supraglacial means, as discussed in the preceding chapters. It was followed by accumulation of sediments beyond Ganshibal area, suggesting the maximum extension of glaciers during the third major glacial advance. This advance removed and transported most of the depositional features produced during the earlier two advances and retreats.

The glacial deposits of the Kashmir valley and the study region have been identified by a number of workers but very little research has been conducted on the morphology and genesis of the different geomorphic features.[1, 2, 3] On the basis of the distribution of the end moraines the glacial history of the East and the West Liddar has been reconstructed.[4, 5] The first systematic study of the moraines has been conducted by Ahmad.[6] He has distinguished the moraines of the longitudinal and transverse glaciers of Kashmir Himalaya and of the study region. The present study deals with morphology, origin and distribution of different moraine types of the area and their age distinction. This work, therefore, is an outgrowth of the findings of the earlier works.

DEBRIS VOLUME

Nearly 40 per cent of the glaciated area is covered by depositional features, out of which nearly 70 per cent is covered by moraines alone. The majority of the morainic soil contain subrounded to rounded rock fragments, brought from the valleyhead and valleywalls of the East and the West Liddar. The high degree of rounding exhibited by rock particles might have resulted from the long-distance movement of sediments in subglacial and englacial position.

GLACIAL DEPOSITS

The major part of the glaciated section of the valley is covered with moraines arranged in ridges, which are approximately parallel to the sides of the valleys. Those ridges which occur in valley basins are rounded. The morainic ridges stretch from Ganshibal terminus to Kolahoi and Shishram snouts. Their continuity is broken at several places by transverse glacial deposits in the East and the West Liddar valleys. In the hanging valley the moraines extend along the hanging walls of the valley. The distribution of the moraines reveal a close relationship between the landform orientation and ice flow direction. The depositional landforms have lineated topography consistently aligned either parallel or transverse to iceflow direction (Fig. 74). The majority of moraines are comprised of unstratified tills. The study shows typical distribution of tills in the moraines. The four principal groups of identified moraines are terminal (end), recessional, lateral and ground. They have been observed at different locales of the valley (Fig. 73). The distribution of the moraines are not simple, most of the moraines are narrow. At certain places, in Shishnag, Vaobal, Kolahoi Ganj, Liddarwat, Zaijpal and Pahalgam, two types of moraines converge to give it a complex form (Fig. 74). At these sites the lateral moraine either joins with the terminal moraine or the ground moraine. Due to this pattern the morphology of moraines has changed appreciably.[7]

Lateral Moraines

Lateral moraines are linear ridges produced from the dumping of till with allied outwash deposits in the trough between the glacier and valleysides. They are one of the commonest depositional features of the region and are aligned parallel to the margin of the East and the West Liddar. They are straight or slightly arcuate. The moraines join their counterpart of hanging valleys at acute or at right angle, such

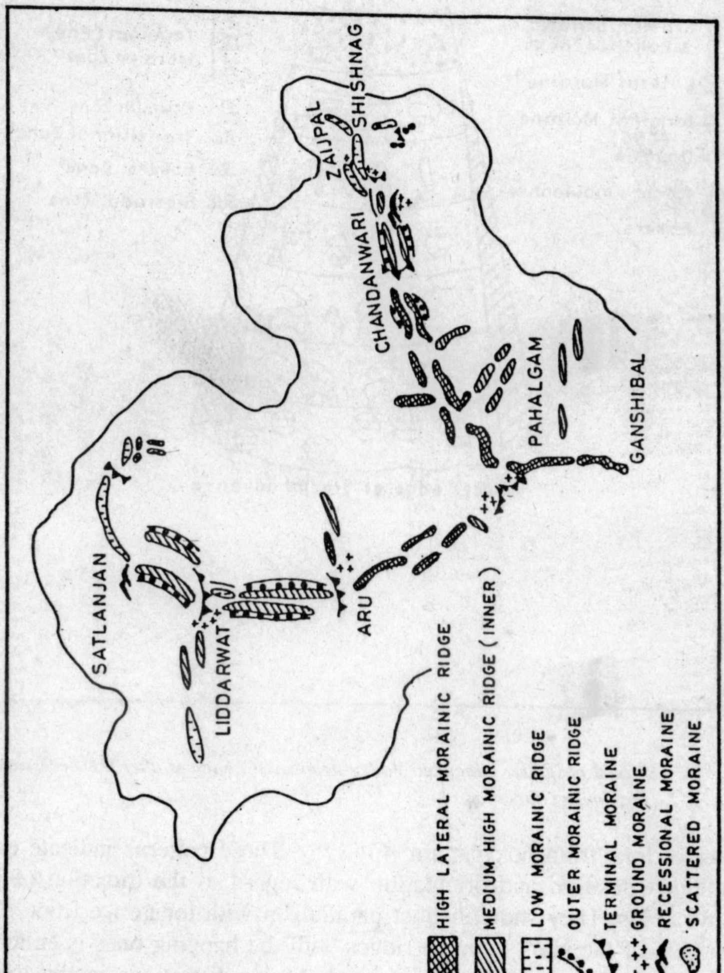

Fig. 73: *Distribution of the Glacial Depositional Features in the Liddar Valley*

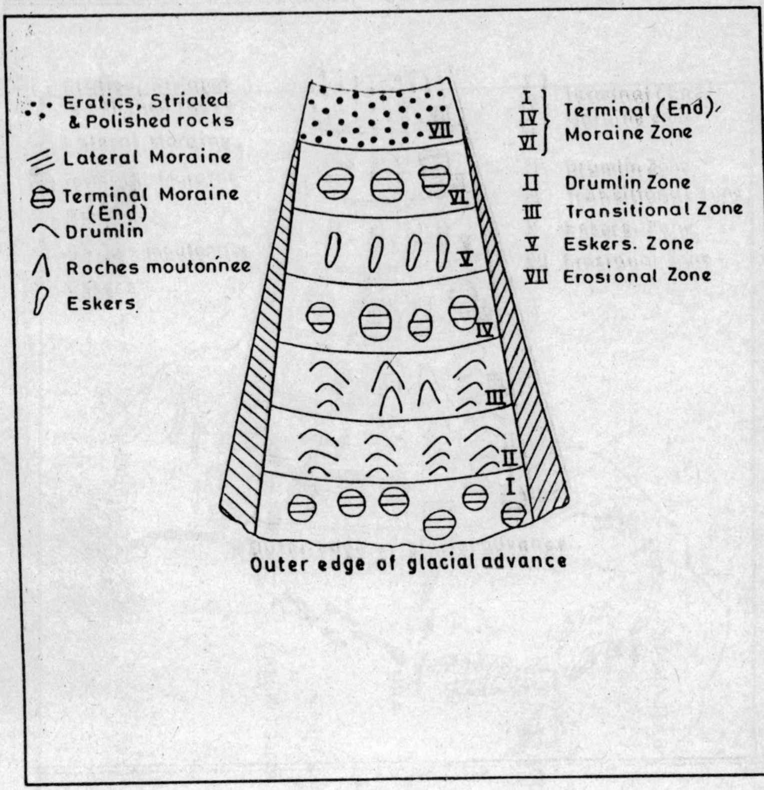

Fig. 74: *Model of Liddar Glaciated Valley-deposition Features after the Recession of Maximum Glaciation*

intersections from box pattern (Fig. 75). These patterns indicate controlled deposition and orientation with regard to the direction of ice movements. They show distinct parallelism with lobate ice front. The junction of the main moraine ridges with the hanging ones is either at the same level or it is superimposed. Major distortions in the lower ridges have been observed, suggesting that the material of the upper hanging moraines while moving have disturbed appreciably the lower main valley morainic ridges. The moraines show distinct gradation in their height from valley bottom to valleyhead. They gradually lose their height from 200 to 20 metres. Their lengths and breadths vary from few metres to 7 kilometres and 50 metres to 2 kilometres respectively. The moraine ridges stretch from Ganshibal to Shishnag and Pahalgam to Kolahoi. Their continuity is broken by hummocky, transverse moraines and spurlines. The morainic ridges of the region have been classified in the field into three groups; high, medium high, and low on the basis of their height and morphological changes (Fig. 73).

There is a distinct uniformity in height and spacing of the moraines in different sections of the valley. The low moraine ridges are found near the valleyhead and high and bulky moraines are located the farthest from the valley. The lateral moraines are primarily composed of thick till containing rounded and subrounded boulders which have been transported subglacially and englacially.

High Lateral Moraines

The high lateral morainic ridges extend along the edges of the East and

Fig. 75: *The lateral morainic ridge (A) near Satlanjan joins the counterpart of Basmai hanging valley (B) at right angle. In the foreground there are heaps of end moraine*

the West Liddar rivers from Ganshibal to Chandanwari and from Pahalgam to Aru respectively. Their continuity is broken at several places by spurlines, ravines and alluvial fans (Fig. 73). They are nearly 100 to 200 metres high from the valleybed and have slope of 15° to 35°. They are large and massive near Pahalgam, Aru and Chandanwari.

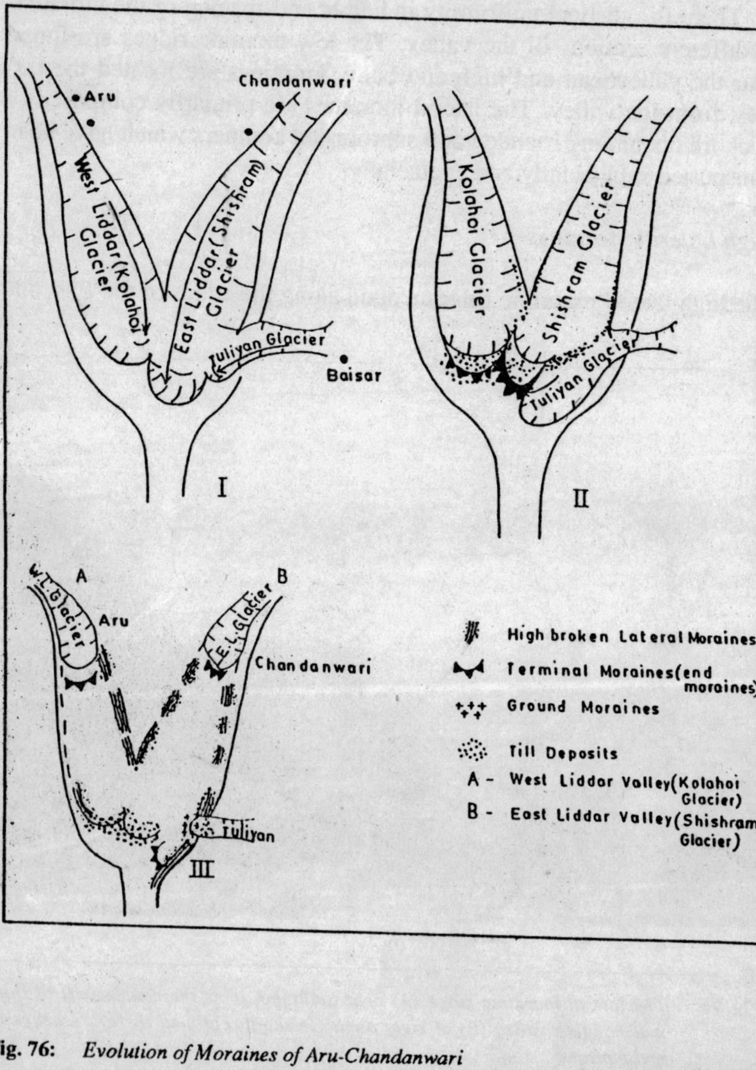

Fig. 76: *Evolution of Moraines of Aru-Chandanwari*

East Liddar Valley: The lateral morainic ridges are preserved as embankments along the eastern margin of the East Liddar between Ganshibal and Pahalgam church (Fig. 77A). The moraines stretch further northward up to Manipal in a slightly arcuate fashion along the western margin of the river, covering a distance of 3 kilometres. At an altitude of 2250 metres near Rainspal the moraine walls are developed on either side of river, and they extend up to Chansan for a distance of 1.5 kilometres. At Chansan they continue northward along the eastern margin of the river up to Nakhpather and Gagrwan spurline (2400 metres) areas for a distance of 2 kilometres. The lateral morainic ridges again reappear in the main valley at an altitude of 2600 metres near Tramakzan. They extend further northward up to Chandanwari along the western margin of the main valley (Fig. 76). In the hanging valleys of Manipal, Dudhal, Wokhbal and Tuliyan the morainic ridges are well preserved along both sides of their tributary channels. In Tuliyan the lateral morainic ridges commence from troughhead and extend up to Baisar, covering a distance of nearly 4 kilometres (Fig. 77B). In Dudhal hanging valley morainic walls have been observed along both the sides of the river over a distance of 2.5 kilometres. The upper end of the moraine is low and articulates with cliff. The height increases to 150 metres towards the terminal end.

West Liddar Valley: The lateral morainic ridges are preserved as an embankment along the western margin of the West Liddar river. It stretches between Pahalgam and Wowurgasu over a distance of 2 kilometres. The moraines are nearly 100 to 150 metres high, 200 to 400 metres wide and have steep slope (25° to 35°) (Fig. 78). At an altitude of 2250 metres, the moraine curves eastward and extends northwards up to Gurkum. Their height gradually increases to 150 metres. At Gurkum they are joined by two hanging lateral moraines originating from Yarnar area (Fig. 73). At this site the moraines of the main valley are breached by alluvial fan. The moraine further extends northwards up to Aru along the eastern margins of the main valley. They are large and massive, standing nearly 150 metres high from the river beds, and are nearly 2 kilometres wide.

Medium High Moraines

The medium high lateral morainic ridges occur between Chandanwari and Zaijpal and Aru and Liddarwat in the East and West Liddar respectively. The ridges are nearly 60 to 150 metres high and are more

Fig. 77A: *Stereoscopic View of High Lateral Morainic Ridge of Ganshibal-Pharsulan Section (East Liddar). The Scarcity of boulders and smoothness of form are suggestive of its very old age (Early Ganshibal stade)*

Fig. 77B: *Old Lateral Moraine of Tuliyan hanGing valley having subdued smooth upper surface and rounded summit*

Fig. 78: *Stereoscopic View of High Morainic Ridge near Mundlun (West Liddar valley) of Early Ganshibal Stade. The scarcity of boulders and absence of terminal moraine are typical*

continuous in extension than the high and the low lateral morainic ridges. The moraines are truncated near Pissu Ghati and Kalwan by spur line. In the vicinity of Chandanwari the medium high morainic ridge run parallel to high lateral morainic ridge for nearly half a kilometre, while in the Zaijpal and the Liddarwat basins two medium high morainic ridges occur adjacent to one another.

East Liddar Valley: The medium high lateral moraines have been identified in the Zaijpal basin. They extend between Pissu top and Nagkut (Burzulkut) along the western margin of the East Liddar river, between the altitudes of 2900 and 3450 metres (Fig. 79B). Two morainic ridges have been identified. They extend parallel to each other. The outer ridge is nearly 150 metres high, while the inner ridge is 75 to 125 metres high (Fig. 80). The morainic ridges are 6 to 10 kilometres long, 2 to 3 kilometres wide and have slopes of 25° to 40°. At an altitude of 3200 metres the lateral moraine curls westwards near Burzukut. At this site two hanging lateral morainic ridges extend to the east of the East Liddar river from Natsar and Sonsar confluence. The moraines extend as hanging walls for a distance of one kilometre (Fig. 73).

West Liddar Valley: The medium high lateral moraines are well developed in the middle section of the West Liddar valley. They stretch along both sides of the river in the two ridges. There are two ridges, one laying inside the other. The outer ridge is nearly 125 metres high and the inner 60 to 100 metres high. The morainic ridges stretch between Kalwan and Liddarwat as ribbons (Fig. 81B). They are 6 kilometres long, nearly 2 kilometres wide and their slope ranges between 25° and 35°. At Gurmarg the inner morainic ridge coalesces with the kame terrace at very steep angle (60°). In the vicinity of Minimarg the morainic ridges are breached along the western end by alluvial fan, while on the eastern side they are slightly arcuate. At Liddarwat two hanging walls of lateral moraines of Tarsar valley join the terminus of medium high moraine at an acute angle in northwest direction at an altitude of 3200 metres (Fig. 81B).

Low Moraines

Low morainic ridges have been identified in the higher parts of the valley near the troughheads of the East and the West Liddar. The moraines are 25 to 50 metres high from the river bed and are 2 to 4 kilometres long. They are very steep. Nearly 15 moraine heaps lie along the inner side of the low moraine ridge.

East Liddar Valley: In the East Liddar valley the low morainic ridge has a curved form. It extends from Shishnag to Vaojan along the western margin of Shishnag valley. A series of small moraine heaps lies along the edge of the present stream trough from Vaojan to Shishnag. Some of the fresh moraines are truncated at the top (Fig. 82A). In the northwest of Shishnag four small lateral moraines are superimposed on the low morainic ridge. They extend from Gulol Gali site at an altitude of 3800 metres towards the main Shishnag valley. At an altitude of 3650 metres, a massive low lateral morainic ridge trends away west and forms lateral bulge as it moves southwest at an altitude of 3550 metres at the mouthing line of Shishnag cirque, like a low embankment.

West Liddar Valley: The low moraine ridge extends from Satlanjan to Kolahoi Ganj area along the western margin of the West Liddar river. The morainic ridge is nearly 4 kilometres long, 250 to 400 metres wide and has very steep slope. The low moraine ridge forms a lateral bulge at an altitude of 3200 metres near the waterfall and moves northeast. Near the Kolahoi Ganj site series of small lateral moraines have been observed adjacent to low lateral morainic ridge (Fig. 82B). Two small hanging lateral moraines extend from the lip of the Hoksar and Dudnag cirque at an altitude of 3600 metres (Fig. 83).

Field Differentiation of Moraines

The lateral moraines of the study region are primarily comprised of rock fragments of volcanics, calcareous sandstones, slates and limestone, ranging in size from clay to boulders. The majority of the moraines contain more than 55 per cent by weight the coarser material (less than 0.1 millimetre), which acts as mild cementing matrix. In the morainic soil the frequency of coarser particles and boulders per unit area and of boulder abundance increase with altitude. Near the valleyhead of the glacial mountains the boulder abundance is maximum; therefore, the material forming most of the moraines at these sites is mainly gravel and boulders with some fine material acting as a rather weak matrix.

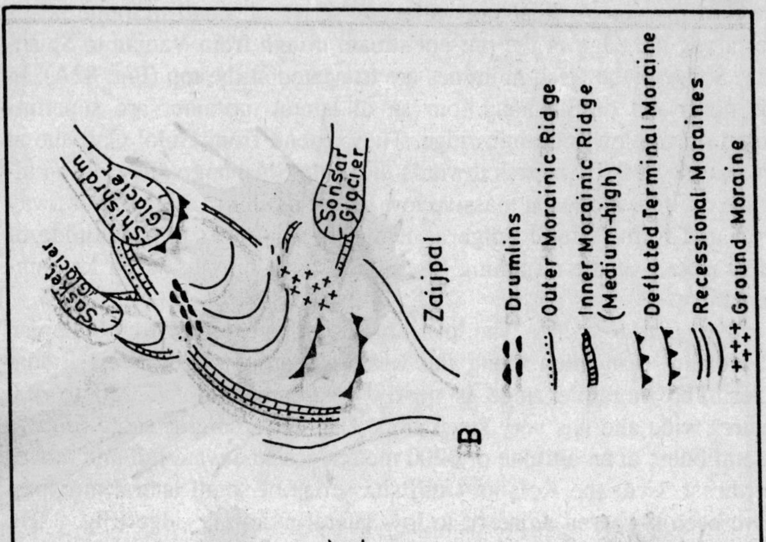

Fig. 79: *Evolution of Moraines of Zaijpal Basin*

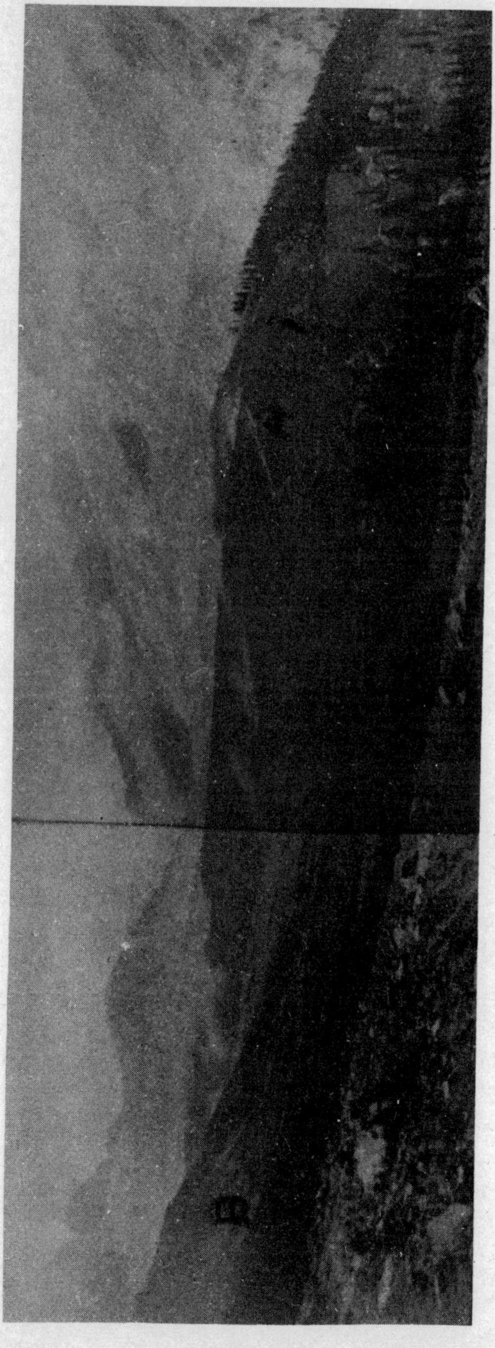

Fig. 80: *Stereoscopic View of Medium High Lateral Morainic Ridges of Zaijpal Basin. Outer ridge (A) (the oldest) is nearly 150 metres high and inner ridge (B) (Middle Ganshibal stade) is 75 metres high, tree line marks the contact.*

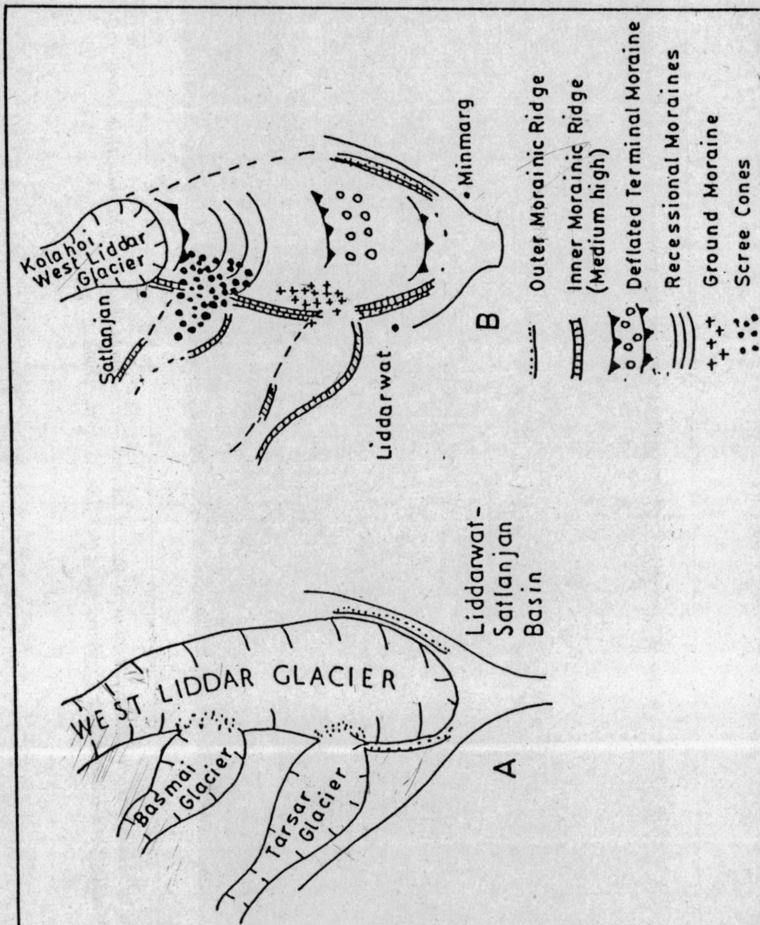
Kolahoi West Liddar Glacier
Satlanjan
Liddarwat
Minmarg
B
Outer Morainic Ridge
Inner Morainic Ridge (Medium high)
Deflated Terminal Moraine
Recessional Moraines
Ground Moraine
Scree Cones
WEST LIDDAR GLACIER
Basmai Glacier
Tarsar Glacier
Liddarwat-Satlanjan Basin
A

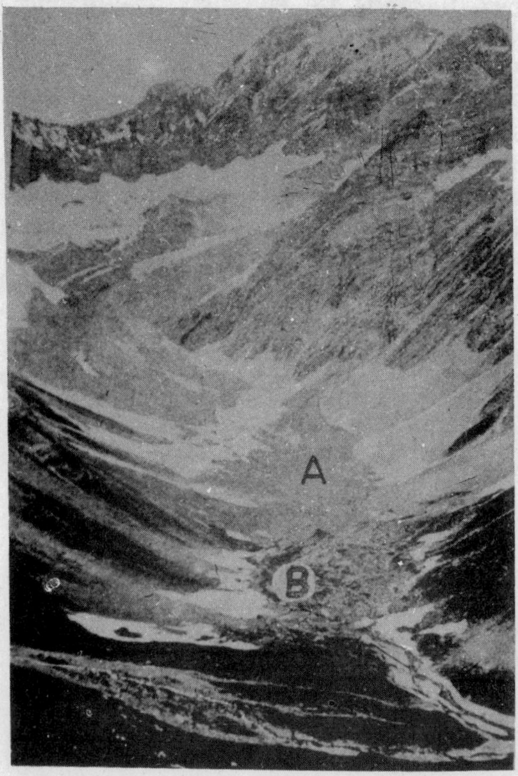

Fig. 82A: *Low Lateral Moraines (A) of Recent Origin Extending out from Shishnag Basin. In the foreground there are heaps of end moraines (B)*

The field differentiation of the lateral moraines has been made on the basis of geometry, morphology of morainic soil, frequency of surface boulders and degree of weathering displayed by those of the selected soil.[8] The assumption that moraines should be progressively older with increasing distance from the cirque was borne out by a comparative weathering and erosional characteristics[9] (Fig. 84). The above parameters have been treated in different ways by the investigators.[10,11]

Boulder Frequency: In the present study, boulder frequency has been determined by counting all surface boulders of a diameter of 30 centimetres or more within a strip, 25 metres long and 5 metres wide, on the exposed crests of morainic ridges, at 15 sites mentioned in the earlier chapter, following the procedure described by Birman and Sharp.[12, 13] Boulder frequency is generally an inverse function of age owing to progressive breakup and burial of surface stone. Abundance of surface boulders was found to vary widely among the moraines, but not always with respect to difference in relative age. The range of variation frequently reflects compositional contrast among moraines built by glaciers whose source areas lay in different lithological belts.[14] The deposits commonly display nearly 250 to 300 boulders in the low moraines and 50 to 100 boulders in an area of 125 square metres of the high moraines respectively. The morphological distinction between different moraines, made by using the ratio of the stones of sedimentary rocks to metamorphic and volcanic (igneous) ones which are more resistant to weathering, is in itself a good indicator of age.[15]

Fig. 82B: *A Series of Low Lateral Moraines (A) Extending Adjacent to the Low Lateral Morainic Ridge (B) of Kolahoi Ganj Area*

Fig. 83: *Fresh Lateral Moraines (A) Extend out from the Cirque Lip of Hoksar. The Fresh Cirque (B) Appear in the Rear*

Grain-size Relationship: The grain-size distribution in the till samples of moraines was determined by using (Indian standard) sieves. The small pits of 40 centimetres long and 10 centimetres wide were due along the exposed portion of moraines near the crest. The soil horizon of the pit was divided in two halves. The material from each of the pit was filled in a metallic sampler of known volume. Thus two samples were collected at each site. The two samples were separately sieved for five fractions; coarse, medium, find sand, silt and clay (Table 27). The analysis reveals that high moraines have higher content of sand fraction and proportionately lower content of finer fraction. The Chandanwari moraines have proportionately lower clay content (18%) than the Aru (23%) (Table 27). The clay content increases with the elevation. In the low lateral moraines the silt content is proportionately higher than the high and medium high lateral moraines. The texture and colour of the till is inconsistent and varies from place to place.

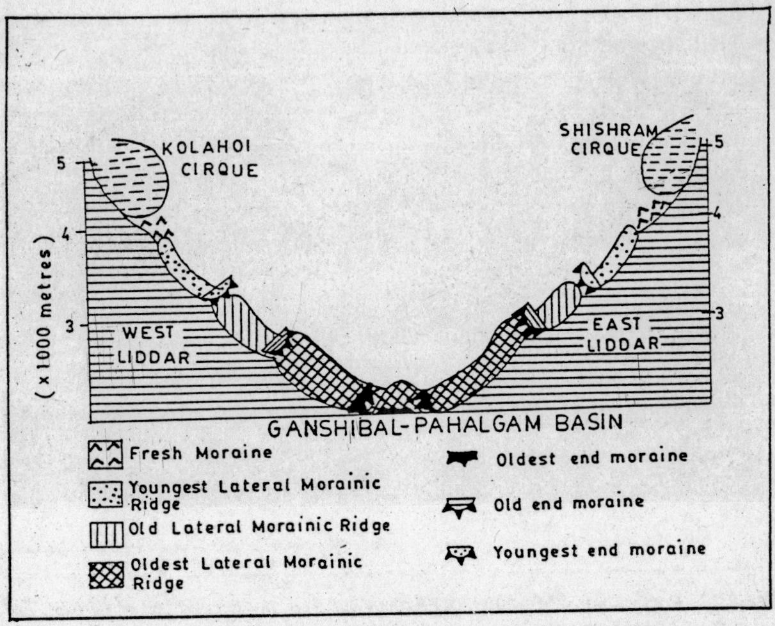

Fig. 84: *Moraines Grow Progressively Older with Increasing Distance from Cirque*

Morphology

The high morainic ridges are very bulky and broad and are nearly 200 metres high. They have subdued upper surface and steep leeward surface. The morainic ridges are mostly fragmented into different lobes in the East Liddar valley as contrasted with continuous lobes of the West Liddar valley. The moraines are less bouldery and contain primarily boulders of volcanic trap and calcareous sandstone. Near Pahalgam the boulders are cemented in brown calcareous soil. There are hardly 75 to 125 large boulders of 30 centimetres diameter in an area of 125 square metres below the morainic crust. The tills comprise primarily of sand fraction (52 to 60%) in which coarser fraction grains predominate (nearly 30%) (Fig. 85). The coarser sand fraction increases near Aru

Table 27: *Particle Size Distribution in Soil Samples of the High, Medium High and Low Moraine Ridges*

Site	Location of section	Position in section down from the top (centimetres)	Nature of moraine	Sand fractions in soil matrix (millimetres)				Silt 0.06 to 0.002mm	Clay below 0.002mm
				Coarse 2-0.5	Medium 0.5-0.25	Fine 0.25-0.06	Total		
1.	Ganshibal	0 - 20	High	28.4	12.8	11.3	52.5	21.7	25.8
		20 - 40	High	28.1	14.9	10.4	53.4	21.3	25.3
2.	Pahalgam	0 - 20	HIgh	28.6	12.5	17.2	58.3	22.6	19.1
		20 - 40	High	29.4	11.4	15.5	56.3	25.7	18.0
3.	Laripora	0 - 20	High	29.9	14.7	15.0	59.6	22.8	17.6
		20 - 40	High	30.1	13.2	15.9	59.2	23.4	17.4
4.	Pharsulan	0 - 20	High	29.8	15.6	16.8	62.2	16.8	21.0
		20 - 40	High	28.7	14.3	14.9	57.9	17.9	24.2
5.	Chandanwari	0 - 20	High	32.7	9.6	10.8	53.1	30.0	16.3
		20 - 40	High	31.8	10.7	11.4	53.9	29.4	16.7
6.	Chandanwari	0 - 20	Med. High	27.6	8.7	10.4	46.7	32.8	20.5
		20 - 40	Med. High	26.8	9.4	10.9	47.1	33.9	19.0
7.	Zaijpal	0 - 20	Med. High	30.4	15.7	7.6	53.7	32.4	13.9
		20 - 40	Med. High	32.8	17.7	12.6	63.1	30.5	6.4

(Contd.)

Table 27: (*Contd.*)

Site	Location of Section	Position in section down from the top (centimetres)	Nature of moraine	Sand fractions in soil matrix (millimetres)				Silt 0.06 to 0.002mm	Clay below 0.002mm
				Coarse 2-0.5	Medium 0.5-0.25	Fine 0.25-0.06	Total		
8.	Minmarg	0 - 20	Med. High	33.6	16.5	10.8	60.9	31.4	7.7
		20 - 40	Med. High	35.1	17.1	11.1	63.3	28.8	7.9
9.	Shishnag	0 - 20	Low	38.2	9.4	7.4	55.0	39.4	5.6
		20 - 40	Low	40.1	6.8	4.6	51.5	39.9	8.6
10.	Shishnag	0 - 20	Low	45.4	14.4	9.6	69.4	16.2	14.4
		20 - 40	Low	47.3	14.8	10.4	72.5	14.6	12.9
11.	Kolahoi	0 - 20	Low	48.6	15.9	11.7	76.2	12.4	11.4
		20 - 40	Low	47.4	15.3	12.2	74.9	12.6	12.5
12.	Satlanjan	0 - 20	Low	35.1	14.2	10.8	60.1	32.6	7.3
		20 - 40	Low	34.8	12.7	10.4	57.9	33.3	8.8
13.	Liddarwat	0 - 20	Med. High	38.7	15.8	12.6	67.1	12.5	20.4
		20 - 40	Med. High	36.4	13.8	11.4	61.6	17.3	21.1
14.	Mandlum	0 - 20	Med. High	28.6	17.4	12.6	58.6	20.4	21.0
		20 - 40	Med. High	29.4	16.8	12.1	58.3	20.1	21.6

Source: Field Observations.

and Chandanwari to 34.3 per cent and 32.6 per cent respectively. The lateral moraines near Chandanwari and Aru have nearly the same boulder count, 75 to 125, though there are minor variations in grain-size. The till deposits are weathered up to a depth of 20 to 25 metres below which it is fresh and sound. The medium high morainic ridges are longer (6 kilometres) and narrower than the high lateral moraines. They are located between the two spurlines of the Zaijpal basin (East Liddar) and the Liddarwat basin (West Liddar). The moraines of this group have been separated by a gap into the inner ridge and the outer ridge to which reference has already been made. The outer ridge is higher, up to 150 metres and steeper (45°) than the inner (60 to 100 metres high and 25° to 35° steep). The three recessional moraines lie near the inner lateral morainic ridge at Zaijpal and Liddarwat. The boulders per unit area increase with increase in altitude. Their number ranges from 50 to 200. The grain-size of the till has a higher concentration of coarse, medium and fine sand fractions (60 to 67%) in comparison to clay-silt (33 to 40%) (Fig. 86).

The ratio of sedimentary to igneous and metamorphic rocks is proportionately lower than in the higher morainic ridges. Stone in these moraines are principally of volcanic trap (basalt) with quartzite veins, limestones, and slates in the East Liddar and volcanic trap (andesite), quartzite with greywacke veins, limestone and shale along the West Liddar valley. The greywacke quartzite lithology is located near the outer morainic ridge of the region. Because of its weathering behaviour it can be used in evaluating the relative ages of the moraines. The weathering of the boulders in the soil of the outer ridge of medium high moraine was observed up to the depth of 30 to 35 metres. Below there was no staining and boulders were fresh. In the inner morainic ridge the staining in the boulders was observed up to a depth of 15 metres only.

The outer slopes of the moraines are dissected by gullies. This is best seen in the lateral view from the southwest. Most of the gullies continue to the moraine top. Their width and depth decrease above the bench level.

The low morainic ridges are very steep (45° to 60°) and are truncated at several places by spurline and deflated alluvial fans. They are unvegetated. The low moraine ridges form a continuous loop near Shishnag and Satlanjan along the northwest in the East and the West Liddar valley respectively. The lateral terminal loop plunges steeply downward towards the East and the West Liddar, reaching the level of

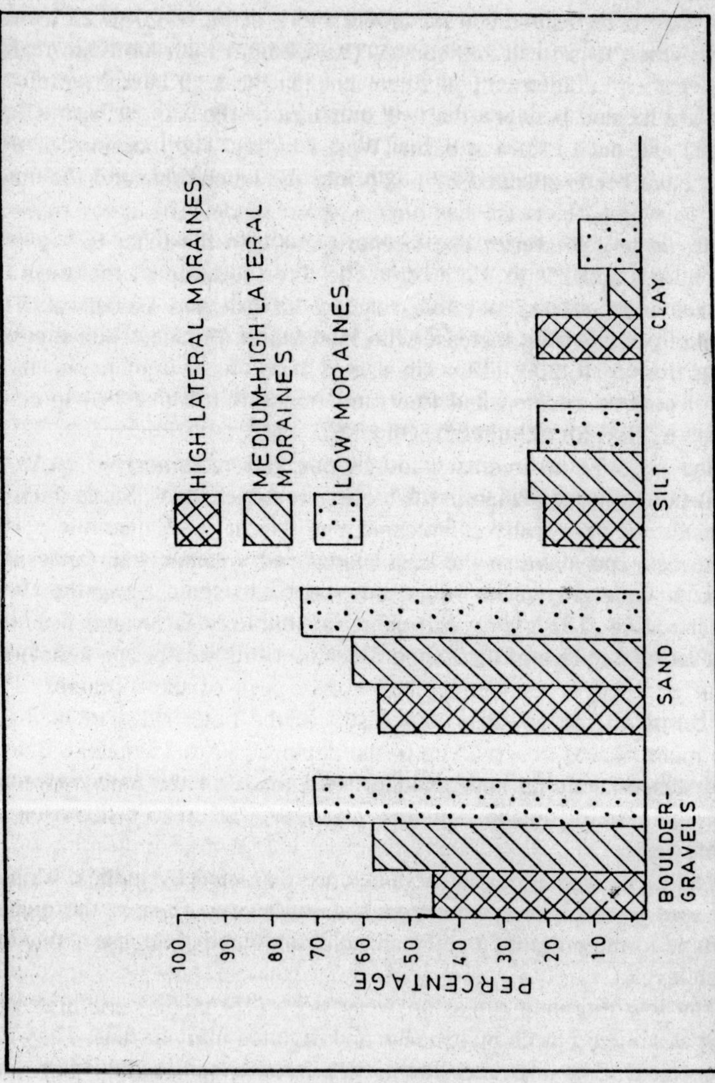

Fig. 85: *Comparison of the particle size distribution in moraines*

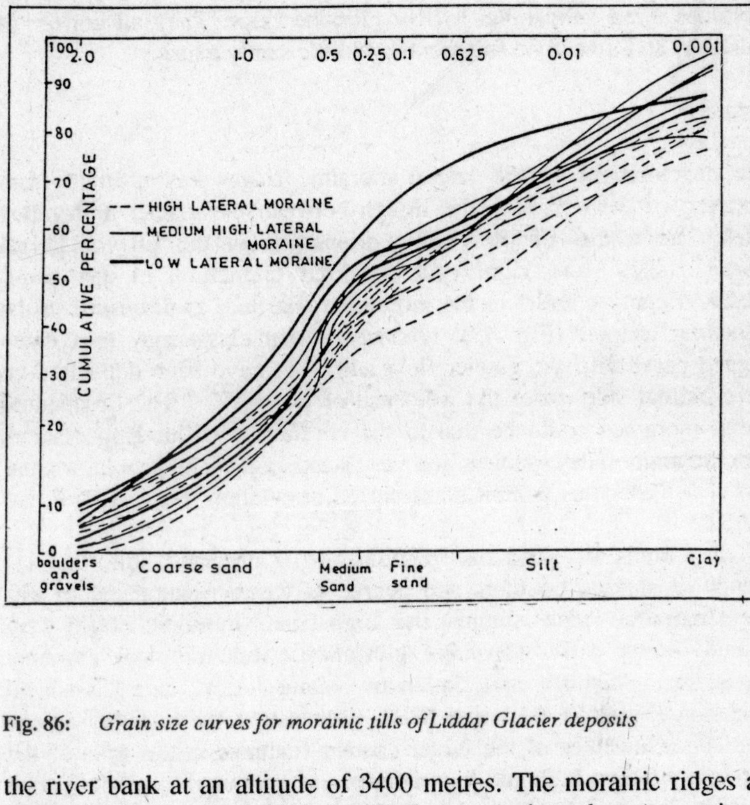

Fig. 86: *Grain size curves for morainic tills of Liddar Glacier deposits*

the river bank at an altitude of 3400 metres. The morainic ridges are nearly 25 to 50 metres high. Inside these moraines there are series of small moraines, 5 to 8 metres high and nearly 100 metres long. The low morainic ridges have relatively abundant boulders on the crest (300 boulders of 30 centimetres or more in an area of 125 square metres) and consist of coarse-grained volcanics and limestone in the Shishnag valley but in Satlanjan-Kolahoi valley it is composed largely of coarse-grained volcanics, shales and limestones. The surface boulders are more plentiful.

The ratio of sedimentary to igneous and metamorphic rocks is relatively low (5.0 to 15.0) in the lower morainic ridges than the medium high morainic ridges. The soil of the moraines contain proportionately lower fraction of finer (24 to 30%) and larger proportion of sand (60 to

76%). The finer fraction primarily comprises of silt. Due to small percentage of finer fraction the soils are unstable and weakly cemented. Near the snout of Kolahoi and Zamtirnag there is multiplicity of small moraines lying subparallel to low moraine ridge. They all consist of volcanics and limestone fragments and little sandy matrix.

Origin

The development of the lateral morainic ridges has resulted from dumping of the debris in the trough between the glacier and valley sides. The results of the pebble orientation in the tills of lateral morainic sites show significant preferred orientation in one hemisphere, dipping parallel to the morainic crest line, as discussed in the preceding chapter (Fig. 72). It suggests that clasts may have been aligned parallel to the glacier flow and must have been deposited on the proximal side under the influence of gravity.[16, 17] The fragmented lateral moraines could be due to the confluence of hanging glaciers with the main valley glaciers at a very steep angle. The pebble orientation near these sites exhibit no preferred orientation (Figs. 72B, F, I, J and L).

The results of grain-size relationship of morainic soil, the frequency of surface boulders and degree of weathering of soil of different morainic ridges suggest that high lateral morainic ridges were formed during early Ganshibal glaciation, while the low morainic ridges near Shishnag and Satlanjan belong to the later Ganshibal glaciation (Figs. 79, 81A and 82B). Both are of the late Pleistocene age. The uniformity of the larger coarser fractions in the soils of the high, the medium high and low moraine are attributed to single glacial advance, but variation in boulder frequency and weathering of soil suggest their difference in relative ages (Figs. 85 and Table 27). The smaller boulder frequency, the highly weathered till and lesser coarser fraction in the morainic soil of the other medium high morainic ridge suggest a marked contrast in age in comparison to other morainic ridges. It is the oldest deposit and older than late Pleistocene. These moraines stretch between Kalwan and Liddarwat and Pissu and Zaijpal in the West and the East Liddar valley respectively (Figs. 79 and 81).

The outer morainic ridge of medium high moraine type could have been formed during the earlier glaciation (second glacial advance). The third glacial advance was the most extensive of all, during which most of the earlier evidences except that of the outer medium high morainic ridge near Liddarwat and Zaijpal were obliterated. This acted as an obstacle to the glacial activity and therefore resulted in the development of subsequent morainic ridge inside of the outer ridge during the retreat of the glaciers (Figs. 79, 80 and 81). The distribution of lateral moraine therefore suggests at least two glacial advances. The last advance was the maximum and it retreated in three phases, each phase produced a set of moraines. The massive moraines produced in the south were formed by the activity of large number of confluent glaciers. The gradual decrease in height towards north is due to smaller glacier thickness (lesser number of confluent glaciers) (Fig. 76).

Ground Moraine

The distribution of ground moraines is highly irregular and scattered. The moraines are preserved only at those sites where the river has cut its valley below the glacial deposit. In the study region the ground moraines have been observed near Pahalgam, Liddarwat and Zaijpal (Fig. 73). The surface expression of the material is dull and monotonous. The surface of the moraines displays scattered large boulders of 5 metres diametre or so, which were dumped by glaciers in the process of melting. They have gently undulating surface which generally increase in altitude from south to north. The surface material, although unassorted, varies from site to site. The material beneath the Pahalgam ground moraines is compact and consists of pebbles and cobbles embedded in silty matrix, suggesting that the ground moraine was probably deposited in cirque lake (Fig. 87B).

The ground moraines of Pahalgam, Liddarwat and Zaijpal are composed of large rounded to subrounded clasts of volcanics with calcareous sandstone, quartzite and limestone respectively.

Source of Material of the Debris with Special Reference to Indicator Boulders

The rocks of volcanic, calcareous sandstone and limestone occupy the core of the anticlinal areas of Pahalgam. The volcanic series occupy the rugged mountainous terrain of the central part of the East Liddar. The calcareous sandstones overlie the volcanic series, followed by limestone beds. The ground moraines of Pahalgam are primarily composed of volcanics and calcareous sandstone. The boulders of volcanics are universally distributed. Their proportion is dependent on quantitative rather than qualitative record. The calcareous sandstone has restricted outcrops between Chandanwari and Laripora. The source material of ground moraine suggests that the ice mass has its source near Chandanwari, as the glacier from the higher altitude has crossed the outcrops of volcanic traps and calcareous sandstone only.

The ground moraine at Liddarwat is primarily composed of the boulders of volcanics and quartzites along the sandy clay matrix. The proportion of quartzitic boulders is more than the volcanic and the boulders are mostly angular. The source region of the quartzitic boulders is located at Tarsar and Basmai where outcrops of quartzites are located. They are four kilometres away from Liddarwat. The angularity of the boulders of the ground moraine is thus due to nearness of source area.

At Zaijpal the source region of the ground moraines is near Shishnag where limestone outcrops are localised.

Fig. 87A: *The Ground Moraine (A) of Tarsar Glacier near Liddarwat*

Fig. 87B: *Talus-filled Pahalgam Cirque Lake by Ground Moraine (A). The varved silty clays are overlain by boulders. (Note) the inclined varved clays reflecting tectonic tilting*

Morphology

The ground moraines at Pahalgam are highly undulating. They are nearly 500 metres long, 400 metres wide and 4 to 10 metres high from the river bed. The pebble and boulder orientation show wide dispersion of values and do not reveal any dominant orientation parallel to the direction of the ice movement. The ground moraine at Liddarwat is the largest one, nearly 2 kilometres long, 750 metres wide and nearly 12 metres high. It increases in height towards north west. It is dominated by coarser material. The pebble and boulder orientation is highly dispersed, similar to that of Pahalgam. The ground moraine at Zaijpal is comparatively smaller and the debris here have larger percentage of particles which stabilised the matrix. The pebbles and boulders show markedly dominant orientation parallel to the direction of the ice movement (Fig. 72).

Origin

The dominant local character of the ground moraines is the matrix of the boulder-clay which suggests their origin to englacial and subglacial processes. The Pahalgam and Liddarwat moraines have been formed from combined ice movement direction, whereas the Zaijpal moraines have evolved from one single ice movement.

Terminal or End Moraines

The end moraine is one of the important geomorphic features formed by the movement of the glacier snout. The form and size are mainly related to ice movement, rate of surface ablation and volume of material contained in the ice. The moraine is used as a tool to delimit the outline of the former valley glacier as it is formed by the dumping of the debris at successive heights by the advancing and retreating glacier. The majority of the end moraines identified in the field were partly destroyed by Liddar river through widening of its flood plains. They have been partly obscured by later fluvio-glacial deposition. Some parts of the end moraines near Pahalgam have been obscured by the overriding ice and are not as prominent as they once might have been. Therefore, belts of small knolls, near Pahalgam, which now have only weak topographic expression might have larger significance in the past. The overriding of moraines have been reported in the Alps and Colorado by numerous workers, who have termed them palimsest moraines.[18, 19, 20] Most of the end moraines observed from Ganshibal to Shishram glacial snouts and from Pahalgam to Kolahoi glacial snouts are still traceable, at selective sites, along the East and the West Liddar valley respectively. The end moraines are markedly located in the trough between the depression and the outermost steep edge of the valley.

The debris of the moraine is primarily composed of volcanics, calcareous sandstones, limestones and shales brought from the headwalls and sidewalls of the East and the West Liddar valley. The boulders and pebbles occupy the major share of the debris. There is varied distribution of tills in the end moraines. At Pahalgam the end moraines have formed gravel and sand lenses between the tills and at Liddarwat and Shishnag the ablated rock fragments have been observed in the tills of the end moraines. The buried, leached zone occurs below the overlying layer.

Distribution

The three terminal loops have been observed near Ganshibal, Kalwan-Chandanwari and Satlanjan-Zaijpal sites of Liddar valley at altitudes of 2000-2150, 2600-2800 and 3200-3400 metres above sea level respectively (Fig. 73). They are most prominent between Aru-Kalwan and Chandanwari-Pissu section of the West and the East Liddar respectively. The end moraines extend along the east bank of Liddar river at an altitude of 2600-2800 metres and are well preserved above the river bed. The three dumps of terminal moraines have been identified in Pahalgam-Ganshibal section at altitudes of 2000, 2050 and 2150 metres in the north, northeast and east of the basin floor. The terminal moraines at higher altitudes near Shishnag and Satlanjan are mostly of deflated type; most of the soil matrix having been removed by axial streams. Near the snouts of Kolahoi and Shishram glaciers a chain of end moraines have been identified. They are primarily comprised of angular rock fragments. The moraines are of ablation types (Fig. 90A).

Morphology

The end moraines have frontal location and they are generally lobate in plan. Two types of moraines are identified in the field, dumped moraine and push moraine. Three dumped moraines have been identified in the Pahalgam-Ganshibal basin. The moraines are marked by heaps of rounded and subrounded boulders embedded in sandy silt matrix. The stony mounds are 5 to 10 metres high, 10 to 20 metres long and located at altitudes of 2000, 2050 and 2150 metres (above sea level) (Fig. 88). In the Aru-Kalwan and Chandanwari Pissu section the end moraines are well preserved as push type end moraine. The moraines at Aru are nearly 75 metres high, 1 kilometre wide and nearly 2 kilometres long. (Fig. 89A). The meltwater streams of the West Liddar have cut through the proximal end. These moraines are preserved because they are resting along and above the river bed.[21] Nearly half a kilometre of the proximal end has been eroded away by the West Liddar river. The push moraines are composed of large boulders, embedded in sand and silt matrix. The top of the push moraine is overlain by 20 centimetres of thick soil profile. The till fabrics show a pebble orientation at right angle to the direction of morainic ridge. Near Satlanjan and Shishnag the end moraines are nearly 1 kilometre long, 1/2 kilometre wide and 25 to 30 metres high from the river bed. The stone

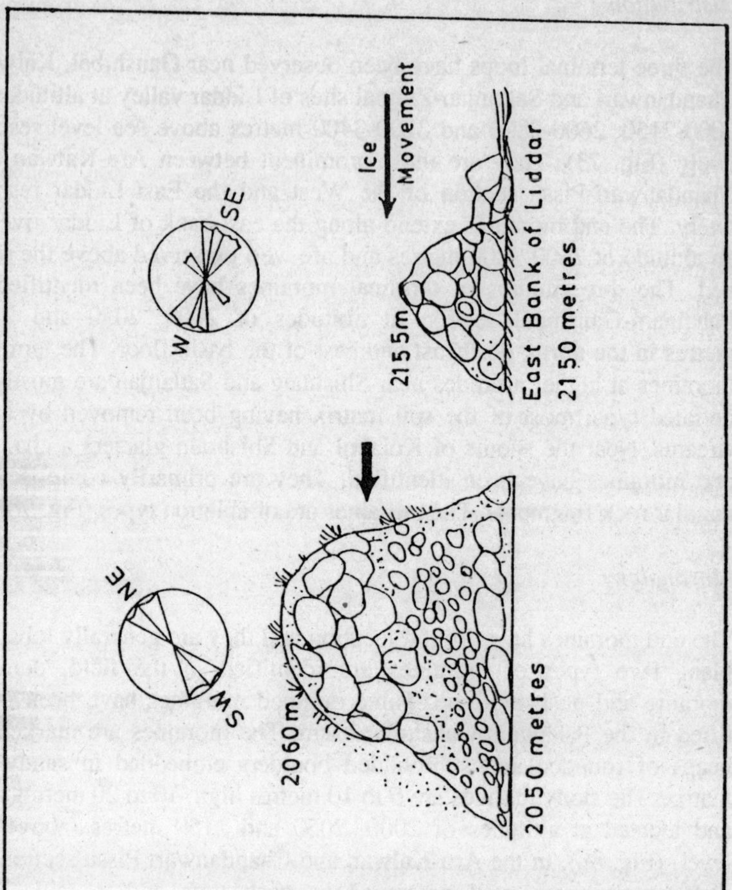

Fig. 88: *Dump End Moraines of Pahalgam Basin*

pavements (large angular blocks) are found on the till top. At Shishnag the Gratbal nala has breached the end moraine.

Origin

The morainic topography in northern part of Liddar valley (East and the West Liddar) is attributed to the third glacial advance followed by three separate glacial episodes. The end moraines identified in Ganshibal-Pahalgam basin is attributed to the third glacial advance or early Ganshibal stade and the end moraines located at Chandanwari-Aru and Zaijpal-Liddarwat are attributable to middle Ganshibal (Chandanwari-Aru interstade) and late Ganshibal (Zaijpal-Liddarwat interstade) glacial episodes.

The three dump moraines located in Pahalgam-Ganshibal basin have their origin in three ice masses coming from three different sources. The dump moraine located near Ganshibal at an altitude of 2000 metres has been formed by the Tuliyan glacier. The dump moraine located at 2150 metres have been deposited by the East Liddar group of glaciers and those located at 2250 metres have their origin in the West Liddar group of glaciers. The mean preferred orientation of the pebbles in the diagram is in north and northeast of the basin (Fig. 88).

The push moraines of Aru and Chandanwari are mainly formed by the West and the East Liddar group of glaciers respectively. The long gentle proximal slope has numerous boulders in sand-silt matrix, the orientation of these boulders and pebble indicate that they are very strongly oriented perpendicular to the crest of the moraine. The till on the proximal side is compact (Fig. 89). The proximal slope is formed by the overriding of Girwar ice front on the main west Liddar glacier. The steep distal slope is loose in texture and lacks in strong till fabrics and is mainly formed out of the material by the sliding down from the proximal slope. The cross-section of the end moraine at Satlanjan and at Shishnag are asymmetric. The long gentle proximal slope has numerous boulders scattered over its surface. The orientation of the boulders, parallel to the moraine front, suggests the origin in the main valley glacier of the East and the West Liddar valleys respectively.

Recessional Moraine

The moraines were apparently formed by the glacier that remained active during the retreat and was characterised by minor advance and temporary halts. The recessional moraines constitute a useful indicator of climatic change. The recessional moraines are usually identified on the basis of the morphological evidence such as the presence of lobes with furrows that show the evidence of flow.[22, 23] Their vertical height ranges from 2 to 15 metres.

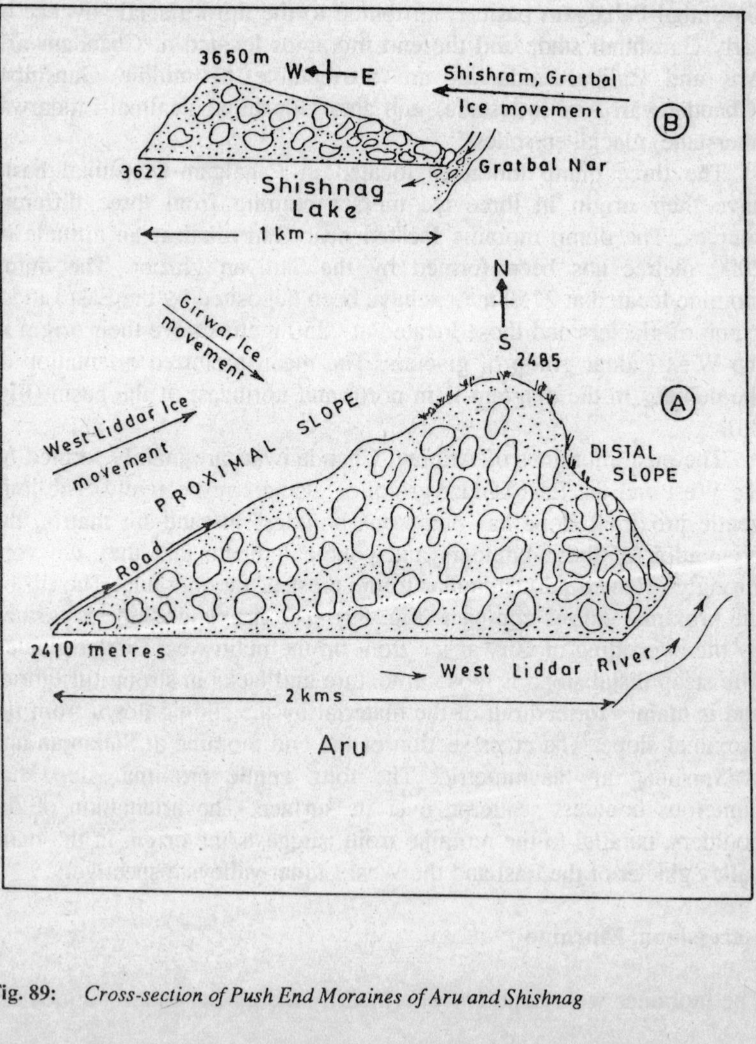

Fig. 89: *Cross-section of Push End Moraines of Aru and Shishnag*

Distribution

The majority of the recessional moraines are confined to one narrow altitudinal belt between 3300 and 3450 metres in the study region. They are located in the centre of Zaijpal and Satlanjan basin of the East and the West Liddar valley. The moraines are piled adjacent to each other and are distributed in stair-like fashion at the basin edge where topographic barrier resisted and slowed ice advance.

Morphology

The most conspicuous morphological feature of the Zaijpal recessional moraine is the set of three lobate bodies of till, arranged in stair-like fashion, closely parallel to one another. The lower stair of the moraine is nearly 1 kilometre long, 300 metres wide and 14 metres high. The two higher stairs are nearly 700 metres long, 450 metres wide and 10 metres high, and they turn northeast along the Zaijpal escarpment. The

Fig. 90A: *End Moraines (A) of Kolahoi Glacier Deposited near Satlanjan*

Fig. 90B: *Recessional Moraine (A) of Shishram Glacier Deposited Northwest of Zaij-pal, shows Lobate form and Furrow Structure (→) made by the Rills*

moraine is slightly convex in the down-valley direction and is covered by furrows indicating flow structure of the glacier (Fig. 90B). The morainic soil has clasts ranging in size from cobbles to boulders of 5 to 6 metres diameter. The diameter of the boulders tends to decrease with decreasing depth. The moraines have undergone considerable post-depositional modification in the northern section where it is crossed by Saskat tributary nala. The moraine consists of kettle holes, kame mounds, and eskers. A giant kettle with diameter of 6 metres has been identified in the moraines. It is covered with weathered limestones.

In the higher reaches of the East Liddar valley at an altitude of 3800 metres near Vaojan pass another recessional moraine has been observed. The moraines have been deposited in three stairs and are piled on the top of each other. They are 4 to 6 metres high and 200 to 350 metres wide and 400 to 600 metres long.

These moraines have been encroached upon in the north by the weathered limestone rock fragments of rock glaciers. A huge kettle has been observed in the third stair of recessional moraine. It is nearly 10 metres long, 6 metres wide and its base is covered with ice.

The Satlanjan recessional moraine has typical lobate form. The moraine consists of two well developed tracts trending northwards. The lower tract is nearly 1½ kilometres long, 500 metres broad and 4 to 6 metre high. The upper tract is 6 metre high and has convex slope. It is composed of a matrix that ranges from clay to boulders and is weakly cemented. The lower tract is mostly covered by angular blocks and is devoid of any sandy matrix. Ahmad mistook the ground moraine of Tarsar deposited on the floor of Liddarwat as recessional moraine (Fig. 90B). Along the retreating glaciers of Shishram and Kolahoi the piles of the boulders occur in front of the snouts. The moraine consists of small mounds dissected by stream. These small mounds are recent recessional moraines deposited by the glaciers.

Recessional Morainic Aspect

There is good correspondence between recessional morainc aspect and microclimatic variables, such as reduced insulation and shading.

From the morainic trends it can be postulated that the Zaijpal and the Satlanjan basins of the East and the West Liddar have been formed by single ice retreat and, therefore, are of nearly the same age. This is indicated by the fact that the recessional moraines in both the basins have similar characteristics: (1) the same lobate trends, (2) they are closely parallel to one another, (3) they have nearly the same width, (4) they have similar topographic expression, and (5) they are trending in the same direction.

NOTES

1. Grinlinton, J.L. (1928). 'Former glaciation of the East Liddar Valley, Kashmir', *Memoir, Geological Survey of India*, No. 49, Part 2, pp. 289-388.
2. Terra, H. De and Paterson, T.T. (1939). *Studies on the Ice Age in India and Associated Human Culture*, Washington D.C.; Carnegie Institute Publication, No. 493, pp. 1-354.
3. Wadia, D.N. (1941). 'Pleistocene Ice Age deposits of Kashmir', *Proceedings of the National Institute of Sciences of India*, Vol. 7, pp. 49-60, ref. pp. 58 and 59.
4. See ref. 1, pp. 380-388.
5. Ahmad, N., and Hashimi, N.H. (1974). 'Glacial history of Kolahoi glacier, Kashmir, India', *Journal of Glaciology*, Vol. 19, No. 68, pp. 279-283, ref. pp. 280-282.

6. Ahmad, N. (1979). 'Morainic deposits in Kashmir Himalayas', *Moraines and Varves: Origin/Genesis/Classification*, Proceedings of an INQA Symposium on *Gensis and Lithology of Quaternary Deposits*, Schluchter, Ch. (Ed.), Rotterdum; A.A. Balkama, pp. 59-63.
7. See ref. 6, p. 61.
8. Sharp, R.P. (1969). 'Semiquantitative differentiation of Glacial Moraines', *Journal of Geology*, Vol. 77, pp. 68-91, ref. p. 69.
9. Porter, S.C. (1970). Quaternary glacial record in Swat Kohistan, West Pakistan', *Bulletin, Geological Society of America*, Vol. 81, pp. 1421-1446, ref. p. 1441.
10. Blackwelder, E. (1931). 'Pleistocene glaciation in the Sierra Nevada and Basin ranges', *Bulletin, Geological Society of America*, Vol. 42, pp. 865-922.
11. Holmes, G.W. and Moss, J.H. (1955). 'Pleistocene Geology of Southwestern Wind River Mountain Wyoming', *Bulletin, Geological Society of America*, Vol. 66, pp. 629-654.
12. Birman, J.H. (1964). 'Glacial Geology across the crest of Sierra Nevada, California', *Geological Society of America*, Special Paper No. 75, pp. 1-80, ref. pp. 16-26.
13. See ref. 8, pp. 69 and 70.
14. See ref. 9, ref. p. 1432.
15. Sharp, R.P. and Birman, J.H. (1963). 'Additions to the classical sequence of Pleistocene glaciations, Sierra Nevada, California', *Bulletin, Geological Society of America*, Vol. 74, pp. 1079-1086.
16. Andrews, J.T. and Smithson, B.B. (1966). 'Till fabrics of the corss-valley moraines of north-central Baffin Island, Northwest territories, Canada, *Bulletin, Geological Society of America*, Vol. 77, pp. 271-290.
17. Wright, H.E. (1957). 'Stone orientation in Wadena drumlin field, Minnesota', *Geografiska Annaler*, Vol. 39, pp. 19-31.
18. White, G.W. (1968). 'Age and Correlation of Pleistocene deposits at Garfield Heights (Cleveland), Ohio', *Bulletin, Geological Society of America*, Vol. 79, pp. 749-752.
19. Ostrem, G. (1961). 'A new approach to end moraine chronology', *Geografiska Annaler*, Vol. 43, pp. 418-419.
20. Norris, S.E. and White, G.W. (1961). 'Hydrological significance of buried valleys in glacial drifts', *U.S. Geological Survey Professional paper*, 424B, pp. 1-78, ref. p. 35.
21. Goldthwait, R.P. (1951). 'Development of end moraines in the east-central Baffin Island', *Journal of Geology*, Vol. 59, pp. 567-577.
22. See ref. 6, pp. 61-62.
23. Totten, S.M. (1969). 'Overridden recessional moraines of north central Ohio', *Bulletin, Geological Society of America*, Vol. 80, pp. 1931-1946.

7

Geomorphological Features of Glacio-fluvial Landscape

The valley glaciers which terminate on the valley basin release large quantities of meltwater that drain broad outwash plain. The rivers which cross these plains erode and rework the previously deposited sediments and redeposit them as distinctive suits of glacio-fluvial lithofacies.[1] This reworking begins under the glacier itself. The landforms resulting from the meltwater deposits retain some of the characteristics of glacial debris. They show a degree of assortment stratification roughly in proportion to the distance they were carried over by streams.[2] The glacio-fluvial clasts have lower degree of rounding because they have not travelled far.[3] The main types of landforms resulting from meltwater deposition are eskers, kame, kame terraces and delta. These are formed at the ice contact situation.

In the Liddar valley during the third glacial advance a majority of the glacio-fluvial landforms disappeared and some of the features such as eskers disappeared as the glacier retreated. They were affected by the glacial process operating in the glacial zone. Many features were formed during deglaciation. Such features are usually concentrated on the lower slopes of valley near valley basin. They are aligned along the major meltwater routes in the lowland basin. Meltwater exists everywhere within and beneath the glacier ice. It is an integral part of the glacier system and generally flows outside the glacier margins.[4 and 5]

MELTWATER STREAMS

Meltwater streams are present throughout the northern part of the Liddar valley from higher altitudes of 4000 metres almost to the Pahalgam basin at 2200 metres. There are nearly 50 meltwater streams, of which 20 have their origin in the present glaciers.

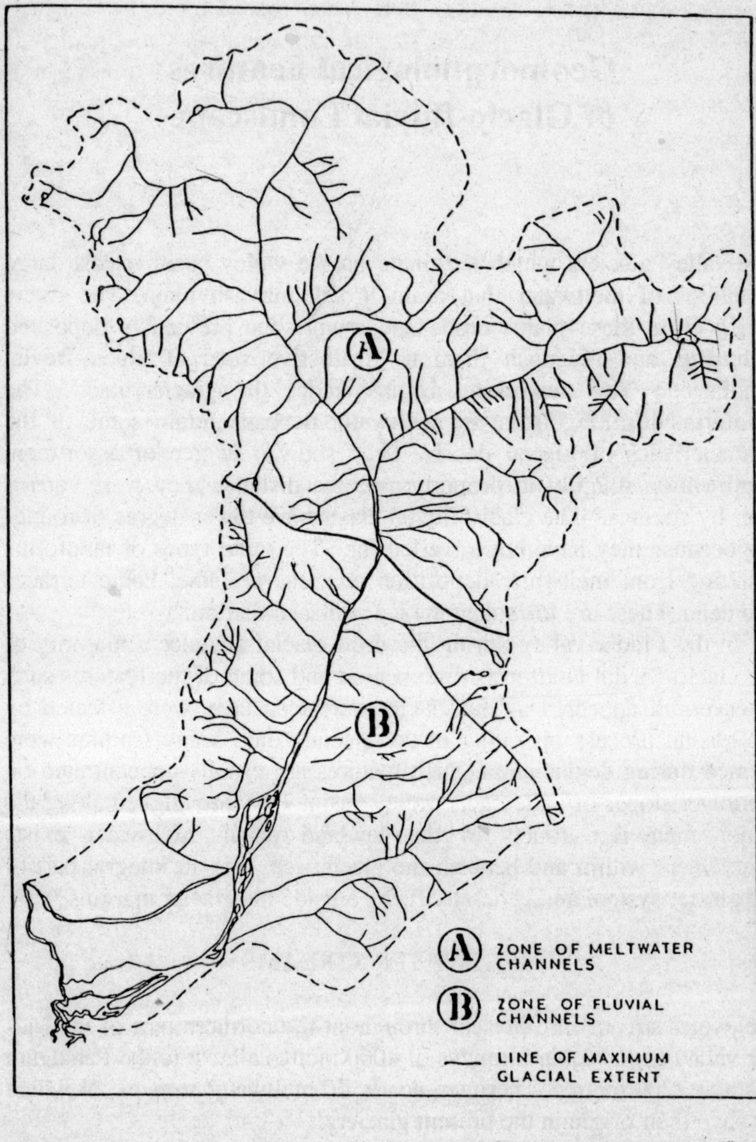

Fig. 91: *Liddar valley: Channel Network*

Distribution

The majority of the meltwater streams are located in the East Liddar valley (Fig. 91). The meltwater streams are tributaries to the main trunk streams of the East and the West Liddar rivers and flow in the northeast-southwest and northwest-southeast directions. The tributary streams join the trunk streams from either side. Most of the streams have developed convex irregularities where the main and the hanging troughs meet. These associations have also been found in Alps and Norwegian meltwater streams.[6,7]

Origin

While observing the existing glaciers one finds water flowing everywhere within and beneath the ablation zone of the glacier. It is flowing either parallel or oblique to the glacier in response to basal ice pressure. The majority of the streams have subglacial drainage. Although many subglacial streams were initiated during the first and second glaciation many have developed during the third maximum state of glaciation. The streams are cut into the interfluves of the East and the West Liddar which retain their preglacial form. Due to the three glaciations, the majority of the streams in the north were incised into the interfluves. They were deepened during the phase of glacio-fluvial erosion. The streams once formed were deepened further by the meltwater of the subsequent glaciation. It has been established that the northern part of the Liddar valley was not structurally controlled but was formed by the glacial and glacio-fluvial processes and Quaternary tectonics. There is no faultline scarp in the northern part of the Liddar valley.[8,9] The presence of fluvial terraces along the rivers clearly suggests that the trunk valleys were formed preglacially as discussed in Chapter 5.[10,11]

Morphology

The morphology of the streams is highly variable. Near the troughheads of the glaciers, the streams tend to be flat-floored with steep sides and are wider than deep. In the middle section of the East and the West Liddar and streams are incised into deep gorges with depth ranging between 75 and 100 metres. The streams are arranged in a system of various size and complexity. Their gradient ranges between 1 in 20 and 1 in 50.

The streams have been divided into two main groups according to their form and location. The first group consists of streams running down the mountain to join accordantly the main flood channel of the East and the West Liddar. The streams of the second group are subparallel to one another. The majority of the glacial streams are located near the troughhead of the northern part of the Liddar valley. They are closely related to the levels of the present stream valleys. They have entrenched themselves into gullies. A majority of them have been cut in the Panjal volcanics and display single intake. The altitude of individual intake varies between 3200-3600 and 3800-4400 metres. Nearly 65 per cent of such streams are located in the East Liddar river. They are quite short (nearly 3 kilometres long) and plunge down the sides of the mountain slopes at very high gradients (they descent from 4400 to 2800 metres). They are the best examples of deep glacial streams. The large dominant streams are more common in narrow gorges, where complex systems are located in broad pre-existing valleyheads and lowlands. The stream systems in these sites have been mainly directed by ice. Clapperton and Sugdon have noticed in the Alps that such streams have two-fold significance.[12, 13] Their existence suggests that the glaciers over the region were warm-based and they provided regional trend in the direction of ice flow.[14]

Subparallel streams are found along the western margin of West Liddar river and southwestern margin of East Liddar river. None of the streams exists above the altitude of 4000 metres, while the majority of them are located below the altitude of 3900 metres (Fig. 91). These streams follow the easiest route. The double intakes are common, the altitude of individual intakes are variable. The places of origin of Son, Chumnai and Gandpathar streams are located at places having altitudinal difference of 100 metres. The Girwar, Manipal and Tuliyan have altitudinal difference of 200 metres. The irregularly spaced parallel streams descend southeastwards between the altitude of 3900 and 2400 metres. The floor of the stream gradient is irregular. They have been deformed by rock and soil creep and are shallower than the first group.

The large dominant stream system (5 to 8 kilometres long) with distinctly different alignments are commonly located at lower altitude. They are found at the former valleyheads throughout the region. Since they are orientated at a small angle to the contours, it is suggested they might have been formed when the glacier had stagnated and thinned to such an extent that free flow of water could take place subglacially down the mountain side to the valleybottom.

ESKERS

Eskers are important features produced by glacio-fluvial processes. They are composed of bedded sand and gravels. They generally show a variety of form and internal composition associated with ice contact glacio-fluvial deposits. The eskers are usually formed in complexes consisting of a number of elongated ridges and hills. The ridges may be straight or sinuous. The concentration of such features is usually found at lower slopes of the mountain. The mechanism of their formation has been given by a number of earth scientists while studying the landforms in the Alps, Norway, and the Rockies.[15, 16, 17, 18] The field observations on the arrangement of eskers with subglacial and supraglacial deposits have helped in a better understanding of the problems associated with ice contact deposits existing in areas long since deglaciated.[19, 20, 21]

Distribution

Two esker complexes have been identified in the study region, Pahalgam and Zaijpal. They cover an area of nearly 10 square kilometres. One esker complex has been observed near Pahalgam. In this complex a long esker ridge is found adjacent to the valley train and lateral moraine ridge. The esker ridge stretches in a curved form at an altitude of 2200 metres along the confluence of the East and the West Liddar rivers. It is nearly 1 kilometre long, 400 metres wide, and 5 to 20 metres high from the river bed (Fig. 92). It is breached by several erosional gaps. The ridge has rounded crest and gentle slope

The second complex of eskers is located near Zaijpal and Shishnag in the East Liddar and at Satlanjan in the West Liddar. The other small esker hills are exposed throughout the area. Most of them are partially buried under outwash and kame terrace deposits. The second complex of eskers consists of small, roughly parallel ridges, 5 to 10 metres, high, 50 to 150 metres long and 50 to 100 metres wide. At Shishnag the top surface of eskers is concordant with the top surface of the kame delta. The kame delta extends along the northern edge of Shishnag lake in east-west direction. In Satlanjan the esker segments are found adjacent to the kame terrace and lateral moraines.

Morphology

The depositional matrix of the Pahalgam esker ridge was investigated for particle size distribution of sediments and shape of pebbles at three road-cut sites near Laripora, Pahalgam race course and Pahalgam village (along the right bank site of the West Liddar). It has been

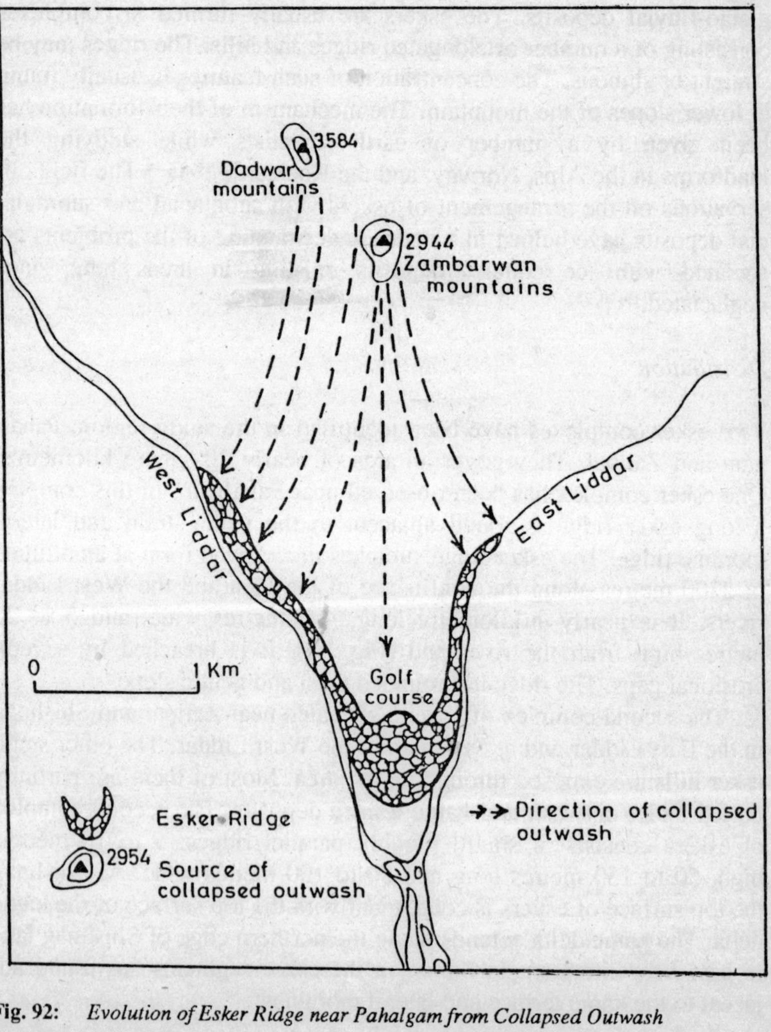

Fig. 92: *Evolution of Esker Ridge near Pahalgam from Collapsed Outwash*

observed that the ridge height decreases from the middle section towards the extreme ends of the ridge near Laripora (East Liddar) and race course area (West Liddar). The darker brown alluvium is found in the upper horizon of the soil profile, the column of soil gradually changes to fine brown below 30 centimetres. The lenses of pebbles are observed below 20 centimetres in sandy matrix. The large column of scratched boulders occur below the pebble bed. The boulders of large and small size show some gradation, they are angular and subangular in form near the central section of the esker ridge (Fig. 93). Below this layer there is a column of large angular pebbles. They are cemented with calcareous rock powder. The lower beds display faint current bedding near Pahalgam at race course area.

Origin

The pebble samples located in the lowest horizon of the soil profile of the Pahalgam esker ridge constitute a homogeneous group. The pebbles of different shapes are cemented in sandy calcareous matrix. The majority of the angular pebbles with calcareous composition suggest their source in neighbouring areas south of Dadwar mountains. The second homogeneous group is formed of angular to subrounded scratched boulders with coarse sand of calcareous composition. This lithological unit is not uniform. It is massive in the centre of the ridge and to its east, but decreases in width towards the west. Possibly it has been deposited by the third glacial advance south of Dadwar mountains towards the Wowurgasu and Laripora in the West and the East Liddar Valley respectively (Fig. 92).[22, 23] The large fault located south of the Pahalgam suggests the possible slumping of sediments from the above sites. The little deformation in the fine structure suggests that deformation has occurred soon after the third glacial advance, because the lower beds show faint current bedding (Fig. 92). The sharp distinction is only formed in the superimposed glacial beds. The origin of the eskers at Shishnag, Zaijpal, Satlanjan and other sites has occurred much later than the third glacial advance. Possibly they were formed later than the Pahalgam esker ridge.

KAME TERRACES

Kames are important landform features of glacio-fluvial deposits which generally show close correspondence with the direction of regional ice movement in the valley glaciers.[24, 25] The valley glaciers often remain active during retreat, inhibiting the survival of subglacial or englacial meltwater and allowing glacio-fluvial landforms to persist only in ice-marginal situation. Also meltwater features built on valley

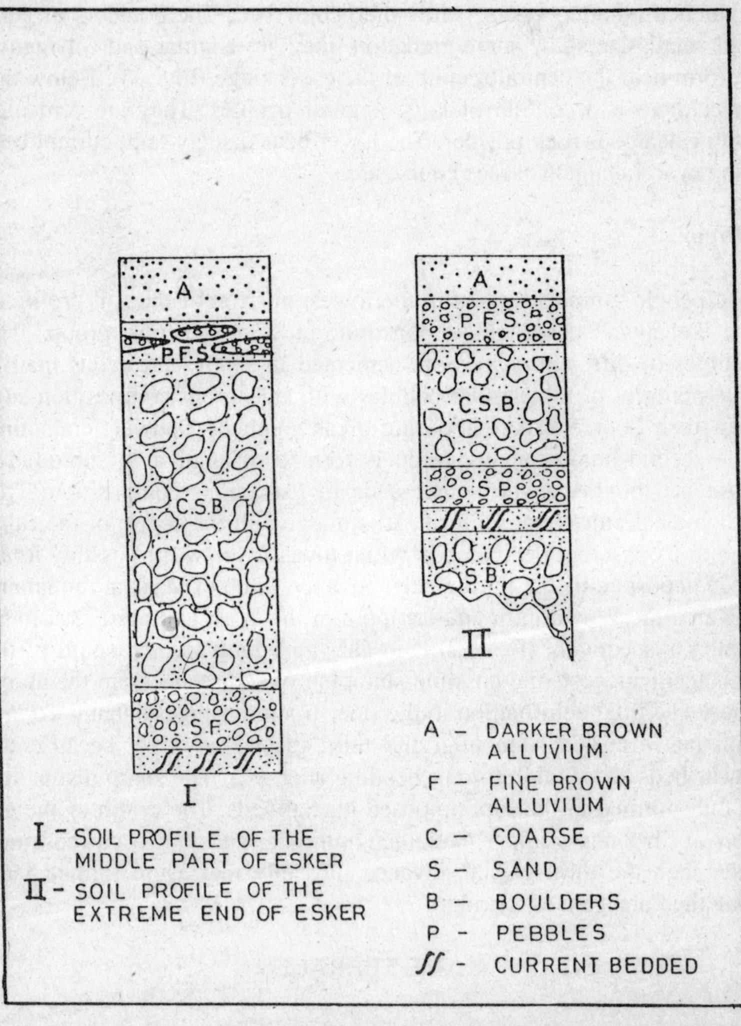

Fig. 93: *Soil Profiles of the Pahalgam Esker Ridge*

floors are partially destroyed by preglacial process. Kames have variable shape because of diverse origins. They are usually composed of poorly assorted water-laid sediments. They also represent the filling of the depression between glacier and the side of trough.[26] These depres sions termed fosse were ultimately graded to terraces. In the study region the kame terraces are less common and are mostly confined in a small altitudinal belt of 2800 to 3000 metres. They are more common in the West Liddar valley than the East Liddar because in the latter the valley glacier, during retreat, remained active, inhibiting the survival of subglacial meltwater forms, which buried the glacio-fluvial features. Nevertheless some of the kames survived along the depression transverse to glacial movement.

Distribution

Kame terraces occur between the altitudes of 2800 and 3000 metres in the East and the West Liddar valleys. They are mainly concentrated in two elevation zones, 2830 to 2850 metres and 2892 to 2905 metres near Drayilyar and Minmarg sites of the West Liddar and at Burzulkut in the East Liddar valley respectively. The kame terrace at Drayilyar is the largest. It is nearly 1 kilometre long and 60 metres wide (Fig. 94). A few of the terraces in the East Liddar are a few hundred metres long and tens of metres wide.

Morphology

In the West Liddar valley the largest kame terraces of Drayilyar and Minmarg have broad flat top in transverse profile. They slope steeply (30°) towards the lateral moraines and dip gently (10°) towards the valley trains. There are several large kettle holes. The stratigraphic sequence of the deposits show that the sediments coarsen vertically upwards. The lowest horizon of the soil profile contains pebble bed of coarse sandy matrix. It is overlain by boulder conglomerate bed having lenses of fine sand and silt. The boulder bed is overlain by a thin layer of yellowish brown sand and silt. The soils are covered by hummocky vegetated surface (Fig. 94). The other kame terraces exposed in small tracts near the East Liddar have also the same soil profile, but there are lithological variations in the boulder conglomerates. The boulder conglomerate bed in the West Liddar is primarily composed of quartzite and volcanics and in the East Liddar limestone and volcanics.

Orientation

The kame terraces of the West Liddar are oriented at 15° to 20° north-northwest. They trend northwest-southeast and are parallel to lateral moraine. Hence they show close correspondence with the direction of ice movement. In the East Liddar the kames are devoid of any consistent orientation with respect of ice movement, but the majority of the small scattered kames have northeast-southwest orientation.

Origin

The position of the kame terraces at the foot of Minmarg and Drayilyar

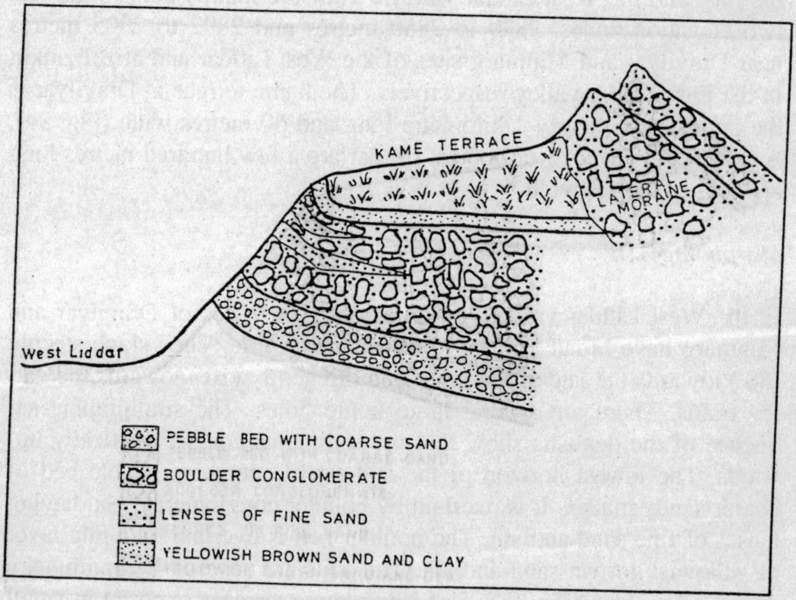

Fig. 94: *Cross-section of Minmarg Kame Terrace*

lateral moraines indicates their close correspondence with the direction of regional ice movement. The stratigraphic sequence of the West Liddar terrace and their flat topped surface suggest their origin in stagnant ice masses and voluminous glacial meltwater deposition.[27, 28] During the glacial retreat the ice of the West Liddar glaciers was trapped in a steep depression of Minmarg and Drayilyar areas. The ice was cut off from the main glacier mass and was pounded in the depression. The retreat of the glacier lead to the accelerated melting of ice. Meltwater from the valleyside and mountain top was pounded in the depression and the level of water slowly rose. The stagnant ice in the depression slowly melted and floated in meltwater like an iceberg. The iceberg dumped the till of the boulder conglomerates over the bed of sedimentary deposits brought by the meltwater channels. The boulder conglomerate was laid over the sand and silt matrix brought from moraine wash. This mechanism of kame terrace development supports the flat-topped formation of the terrace and also reveals the source of the boulder conglomerate in the northwest part near Tarsar and Basmai section of the valley. The parent rock material is the same as that of Tarsar and Basmai quartzite.

The poor kame terrace developments in the East Liddar valley is related to the scattered, shallow depressions and thick ice mass in broad valleys formed by confluent glaciers. This resulted in little stagnation of ice mass. In the Burzulkut area, the blocks of ice were scattered and few scattered small kame terraces were formed. They got till deposits from nearby Shishnag areas. Due to scattered distribution the majority of the kames are devoid of any consistent orientation.

DISTRIBUTION AND MORPHOLOGY OF KAME DELTA

A large kame delta has been identified adjacent to Shishnag lake at an altitude of 3670 metres (Fig. 95). The delta is nearly 1.3 kilometres long, 300 to 500 metres broad and nearly 150 metres high from the Shishnag lake. It has a flat-topped surface which gently slopes to the northwest. The middle section of the delta has steep slope (35° to 50°) towards the lake, indicating the ice contact slope and suggesting the possible level of englacial water-table. Over the top of the delta surface, three esker hills are located at altitudes of 3637, 3648 and 3652 metres respectively.

The stratigraphic section of the delta is primarily composed of two homogeneous groups as indicated by soil profile. The lowest horizon near the Shishnag lake is primarily composed of pebbles and cobbles with coarse sandy matrix. The bed passes upwards into clayey silt bed with cobble-boulder matrix. This bed stretches for 10 metres and covers the major part of the soil profile. The boulders are mainly angular to subangular in form. They are primarily composed of limestone, nearly 60 per cent of them are striated. The boulder bed is overlain by blackish brown silt, rich in humus. Towards the northeast the delta surface is covered by large kettles. The soil profile reveals that soil texture becomes coarser towards the top.

Origin

The huge size of kame delta suggests that such a feature could have

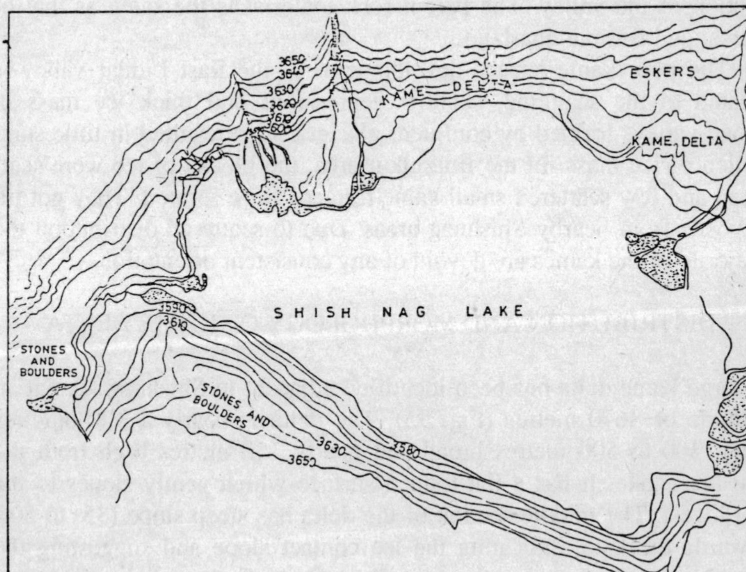

Fig. 95: Kame Deltas Around Shishnag Lake

been formed by thick glacial ice and by large quantity of glacial meltwater. During the third glaciation 8 tributary glaciers converged towards the northwest of the cirque lake and later moved southwest along the East Liddar valley. These 6 tributary glaciers entered the cirque lake from southeast side and moved northwest and other 2 branches came from the east. These :onverged at 10° to 15° north-northwest. Due to the convergence of the glaciers the Shishnag valley was covered of ice. During deglaciation huge quantity of meltwater was formed by streams. They carried heavy load and deposited in deltaic form along the northwestern side of the lake. Due to this process there was possibly a damming of sediments along that side. The glacier broke down into large icebergs by the heat of meltwater. Since the icebergs moved towards the northwest the till was lodged in that direction. By this mechanism the deltaic deposits merged into longer features aligned parallel to ice front.

NOTES

1. Price, R.J (1964). 'Land forms produced by wastage of the Casement glacier southeastern Alaska', *Institute of Polar Studies*, Report No. 9, pp. 1-164.
2. Lee, H.A. (1965). 'Investigation of eskers for mineral exploration', *Geological Survey of Canada*, Vol. 65, No. 14, pp. 1-17, ref. p. 5.
3. King, C.A.M. and Buckley, J. (1968). 'The Analysis of stone size and shape in Arctic environments', *Journal of Sedimentary Petrology*, Vol. 38, pp. 200-214.
4. Shreve, R.L. (1972). 'Movement of Water in Glaciers', *Journal of Glaciology*, Vol. 11, No. 62, pp. 205-214.
5. Stenborg, T. (1969). 'Studies of internal drainage of glaciers', *Geografiska Annaler*, Vol. 51A, No. 1-2, pp. 13-41.
6. Mannerfelt, C.M. (1949). 'Marginal drainage channels as indicators of the gradients of the Quaternary ice caps', *Geografiska Annaler*, Vol. 31, pp. 194-199.
7. Boer, G. De (1949). 'Ice-margin features, Leirbreen, Norway', *Journal of Glaciology*, Vol. 1, pp. 332-336.
8. Bhat, D.K. (1978). 'Geological Observation of Margs in Kashmir Valley, India', *Himalayan Geology*, Vol. 8, Part 2, pp. 769-783, ref. p. 781.
9. Koul, B.L. (1976). 'Structures of Liddar Valley Anticline', *International Seminar on Himalayan Geology*, New Delhi, p. 59.
10. Devendra Pal and Srivastva, V.K. (1982). 'Landform configuration of Karewa floor and its implication on the quaternary sedimentation, Kashmir Himalaya', *Himalaya: Landform and Processes*, Verma, V.K. and Saklani, P.S. (Eds.) New Delhi: Today and Tomorrow Printers and Publishers, pp. 47-75.
11. See ref. 8, p. 773.
12. Sugden, D.E. (1970). 'Landforms of deglaciation in the Cairngorm mountains, Scotland', *Transactions and Papers, Institute of British Geographers*, No. 51,

pp. 201-219.

13. Clapperton, C.M. (1971). 'Location and Origin of Glacial Meltwater Phenomena in the eastern Cheviot hills', *Proceedings, The Yorkshire Geological Society*, Vol. 38, pp. 361-380.
14. Sugden, D.E. and John, B.S. (1976). *Glaciers and Landscape*, London; Edward Ar nold Ltd., pp. 1-376, ref. pp. 281-286.
15. Price, R.J. (1973). *Glacial and Fluvioglacial Landforms*, Edinburgh; Oliver and Boyd, pp. 1-240.
16. Price R.J. (1969). 'Moraines, sandar, kames and eskers, near Breidamerkur-Jokull, Iceland', *Transaction and Papers, Institute of British Geographers* No. 46, pp. 17-43.
17. Suggate, R.P. (1958). 'Late Quaternary deposits of the Christchurch metropolitan area, New Zealand', *New Zealand Journal of Geology and Geophysics*, No. 1, pp. 103-122.
18. Holmssen, G. (1963). 'Glacial deposits of south-eastern Norway', *American Journal of Sciences*, Vol. 261, pp. 880-889.
19. Lewis, W.V. (1949). 'An esker in process of formation, Boverbreen, Jotunheim, 1947', *Journal of Glaciology*, Vol. 1, pp. 314-319.
20. See ref. 3, pp. 1-16.
21. Wisniewski, E. (1973). 'Genesis of Lammi eskers (south Finland)', *Fennia*, Vol. 122, pp. 1-30.
22. Jewtuchowicz, S. (1965). 'Description of Eskers and Kames in Gashamnoya and on Bungerbreen, south of Hornsound, Vestspitsbergen', *Journal of Glaciology*, Vol. 16, No. 41, pp. 719-725.
23. Hellaakoski, A. (1931). 'On the Transportation of Materials in the Esker of Laitila', *Glacial Deposits*, Goldthwait, R.P. (Ed.), U.S.A.; Benchmark Paper 21, pp. 282-311.
24. Niewiarowski, W. (1963). 'Types ot kames occurring within the area of the last glaciation in Poland compared with kames known from other regions', *Report, 6th Conference of International Association of Quaternary Research*, Warsaw, Lodz, Vol. 3, pp. 475-485.
25. Sissons, J.B. (1958b). 'The deglaciation of part of East Lothian', *Transactions and Papers, Institute of British Geographers*, No. 25, pp. 59-78.
26. Holmes, C.D (1947). 'Kames', *American Journal of Sciences*, Vol. 245, pp. 240-249.
27. Mckenzie, G.D. (1969). 'Observations on collapsing kame terrace in Glacier Bay National Monument, South-east Alaska', *Journal of Glaciology*, Vol. 8, pp. 413-426.
28. Jahn, A. (1969). 'Kame terraces in the Studies', *Folia Quaternaria* Vol. 30, pp. 17-20.

8

Fluvial Geomorphological Features

The long stretch of the Liddar valley from the sources of the East and the West Liddar down to their confluence at Nunwun and further to the junction of Liddar with Jhelum at Gur is constituted a fluvial landscape. There is, in it, a preponderance of gorges and terraces in the northern and the southern parts respectively. In the northern parts the two principal tributaries of the Liddar river are cutting their beds vertically at greater rate than laterally. Over a distance of 25 kilometres the rivers are hemmed in by slopes that rise precipitously from the water edge. These processes result in the development of gorges with narrow bottoms and steep sides. Near Ganshibal the Liddar river cuts the last of the inner ridges and the rapid down cutting is followed by leisurely widening and reaggradation. The outer flanks of the Liddar river beyond Ganshibal are covered by beds of coarser sediment such as are found only in the main channel. There sediments have formed the terraces. There is marked correspondence between the distribution of accumulated coarser sediments and the river gorges[1] (Fig. 96).

GORGES

The northern and middle parts of the East and the West Liddar flow through the gorges created by them. The East and West Liddar rivers have a tortuous course in their upper stretches. A majority of the gorges in the study region are cut in the limestone and volcanic formations. In the resistant limestone formation of the Shishnag and Zaijpal localities the East Liddar river has a narrow channel less than 20 metres across and its gorges are deeply incised (Figs. 10A and 10B). Near the localities of Chandanwari and Pharsulan the softer calcareous sandstone rock between the gorge and the East Liddar river has carved broad channel up to 75 metres wide.

Distribution

The East and the West Liddar rivers from the source of their confluence pass through a series of six gorges. The gorges are carved in thick volcanic, limestone and calcareous sandstone formations at the localities of Shishnag, Zaijpal, Pissu Ghati and Pharsulan in the East Liddar and at Liddarwat and Kalwan in the West Liddar respectively (Fig. 96). Just below the confluence of the East and the West Liddar rivers another gorge is found at Ganshibal.

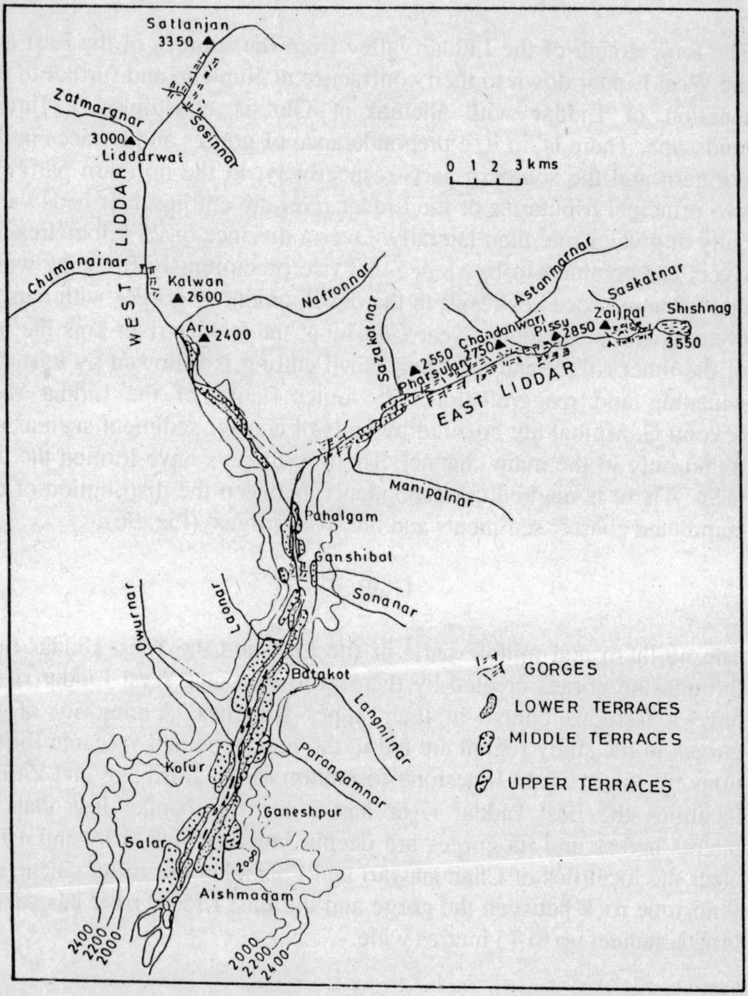

Fig. 96: *Distribution of Gorges and Terraces in the Liddar Valley*

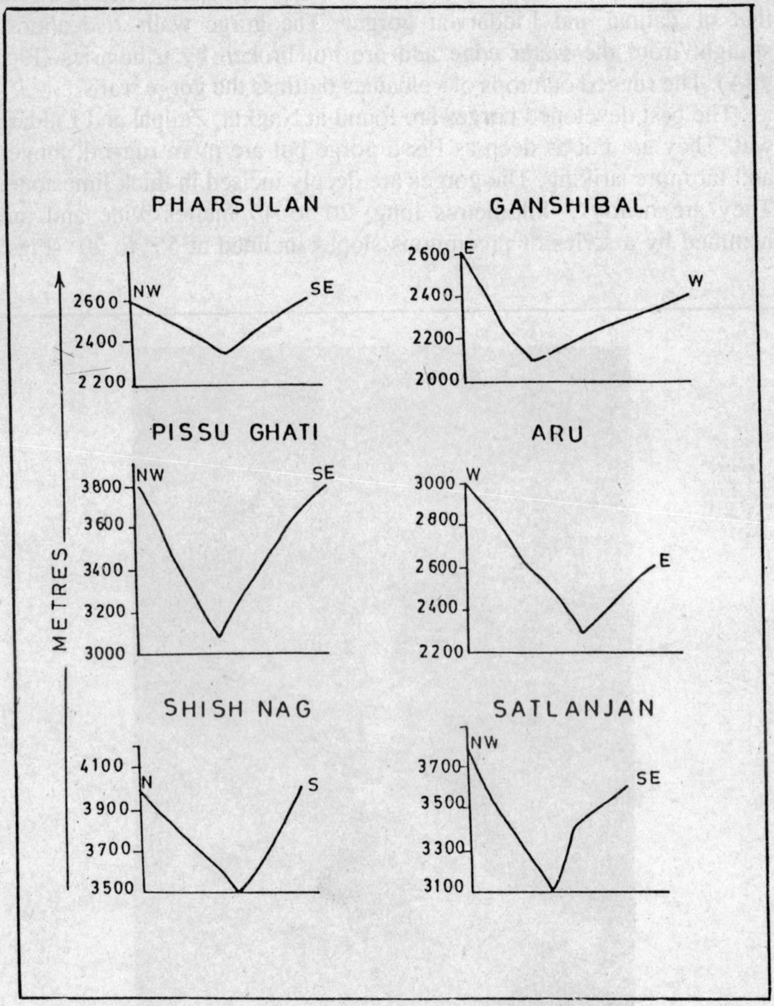

Fig. 97: *Comparative Profiles of Some of the Liddar Gorges*

Morphology

The profiles of the several Liddar gorges are shown in Fig. 97. The deepest of these by far is Pissu gorge. This gorge is nearly 500 metres long, 30 metres wide and 300 metres deep. Its depth is nearly twice that of Zaijpal and Liddarwat gorges. The gorge walls rise almost straight from the water edge and are not broken by tributaries (Fig. 98A). The rugged outcrops of volcanics buttress the gorge walls.

The best developed gorges are found at Nagkut, Zaijpal and Liddarwat. They are not as deep as Pissu gorge but are more rugged, longer and far more striking. The gorges are deeply incised in thick limestone. They are nearly 3 kilometres long, 20 to 40 metres wide and are hemmed by a series of precipitous slopes inclined at 55° to 70° (Figs.

Fig. 98A: *Narrow and Steep Pissu Gorge. The expoliation along steep volcanic walls contribute to sheerness of joints*

99A and 99B). The gorge walls rise almost 59 to 150 metres from the valley bottom and are lined with a large number of hillside tributaries. A number of waterfalls are formed at the junction of side tributaries of Sasket and Basmai with the main channel of the East and the West Liddar respectively. The waterfalls are nearly 20 metres high. They have resulted from topographic unconformity due to capture of side tributaries by the main channel (Fig. 96). The cascades are observed all along the courses of the gorges. The walls, of Zaijpal gorge are pocked by honeycomb like hollows.

Fig. 98B: *Deeply Incised Limestone Gorge of Zaijpal (A). The gorge has been eroded to a depth of 70 metres*

Fig. 98C: *The Well Developed Gorge near Ganshibal Produced by Faulting. The cascades are observed along the course of the gorge*

Pharsulan gorge is the shallowest and stretches between Chandanwari and Pharsulan. The average depth is about 40 metres. It is nearly 2 kilometres long and 50 to 75 metres wide. The gorge walls are made up of calcareous sandstone and slates. Its bottom is covered with apron of large angular talus blocks at many places. The great size of these blocks interrupt the course of the East Liddar. The rock blocks in excess of what East Liddar river can handle in a given time form talus against the gorge walls (Fig. 99B). A small gorge at Ganshibal is nearly 400 metres long, 25 metres wide and 75 metres deep. The western wall of the gorge rises almost straight from the water edge.

The eastern gorge wall is comparatively gentler and is inclined at 30° to 60°. The walls of gorge are composed of volcanics. Since volcanics are highly jointed, the joints have aided in rapid downward cutting by the Liddar river. The gorge has been produced by faulting (Fig. 98C)

Origin

The gorges of the northern part of the Liddar valley were formed due to the slow uplift of the mountain ridges across the paths of the East and the West Liddar river.[2,3] Several geological factors in combination have helped the formation of the gorges. The gorges have resulted partly from slow upheaval and partly from slow tedious process of erosion.[4] The Liddar river drops by more than 700 metres in the gorge country. The steep gradient provides the river with the energy it needs to incise its beds at a rate higher than the rate of the lateral erosion

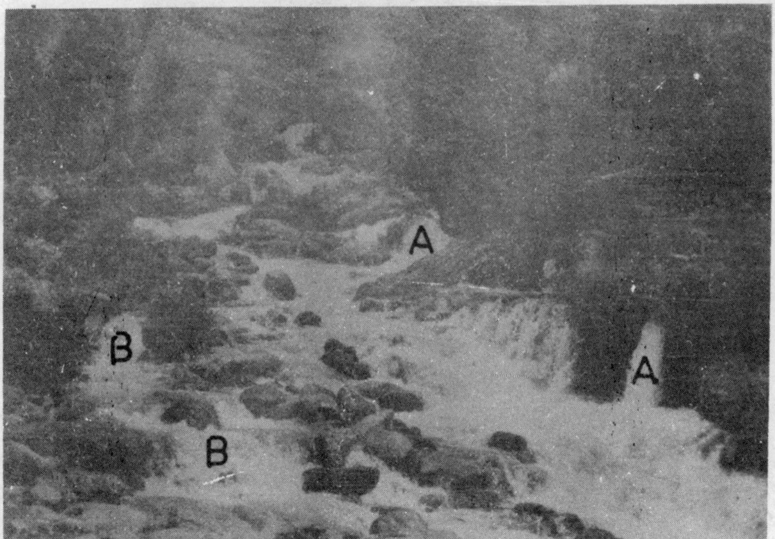

Fig. 99A: *The Shallow Chandanwari Gorge. The gorge walls are made of calcareous sandstones. The waterfalls (A) at right and rapids (B) on the left are 5 and 2 metres high respectively*

(flaring out).[5] The work of the river is seasonal. During the summer the Liddar river has a peak discharge of 50 metres per second through the Ganshibal gorge. At this discharge the river expends energy at the rate of 15 million foot pounds per second, or 1/2 million horse power.[6] Most of this energy is dissipated by the turbulent flow. Part of this energy is expended on scouring the bed of the rivers and the remainder is consumed in transporting the debris. Frost action of course is primarily a winter process. It takes the advantage of such natural weakness in rocks as fractures and joints as discussed in the preceding chapters.

Fig. 99B: *The Talus Boulders (A) in Pharsulan Gorge are being slowly swept Downstream by Floods*

The deep Pissu gorge owes its origin to the cutting of its volcanic walls by the combined flow of the East Liddar and the Sind rivers near Pissu region during Pleistocene times. The grand junction of the two rivers near the Pissu was once facilitated by the capture of Kelnar, a tributary of Sind river, by Astanmargnar, a tributary of East Liddar river. The flow of the two rivers was diverted around the northeast end of Pissu in a remarkable sequence of drainage adjustments (Fig. 100). As a direct consequence, the energy and competence of the East Liddar system were vastly increased.

The Zaijpal and Liddarwat gorges are mostly lithological in character. They are formed in limestone formations near the junction of the large East and the West Liddar rivers with the side streams. The large stream deepens its valley rapidly than the side streams. The main East and the West Liddar incised the limestone more than their side streams. The hard quartzites and slates occurs at the junction of the main stream with the side stream. This has resulted in the formation of the waterfalls.

The broad and shallow gorges near Pharsulan and Aru are also lithological in character. The gorge walls are made of weak and strong rocks of calcareous sandstone and slate. The weak and strong beds in alternation have resulted in formation of broad, shallow gorges with waterfalls which interrupt the course of the river.

The Ganshibal gorge has resulted from faulting. Due to faulting there has been subsidence of volcanic beds along the eastern side of the Liddar river. The subsidence of beds resulted in the lateral shifting of Liddar river followed by deepening of its bed and formation of a gorge.

Terraces

Terraces are flat or gently inclined land surfaces produced by fluvial action. They have been classified into erosional and depositional ones on the basis of their genesis.[7, 8] The depositional terraces are the remanents of palaeoriver floor or floodplains abandoned by parent stream due to shift in geomorphic balance between erosion and deposition. The shift in the geomorphic balance is brought about by changes in climatic and tectonic conditions.[9] The depositional terraces have received maximum attention from earth scientists as they constitute important morphostratigraphic unit, where palaeo time-table of geomorphic processes are faithfully recorded.[10, 11] Lot of work has been conducted on the terraces of Himalayan rivers and it has been

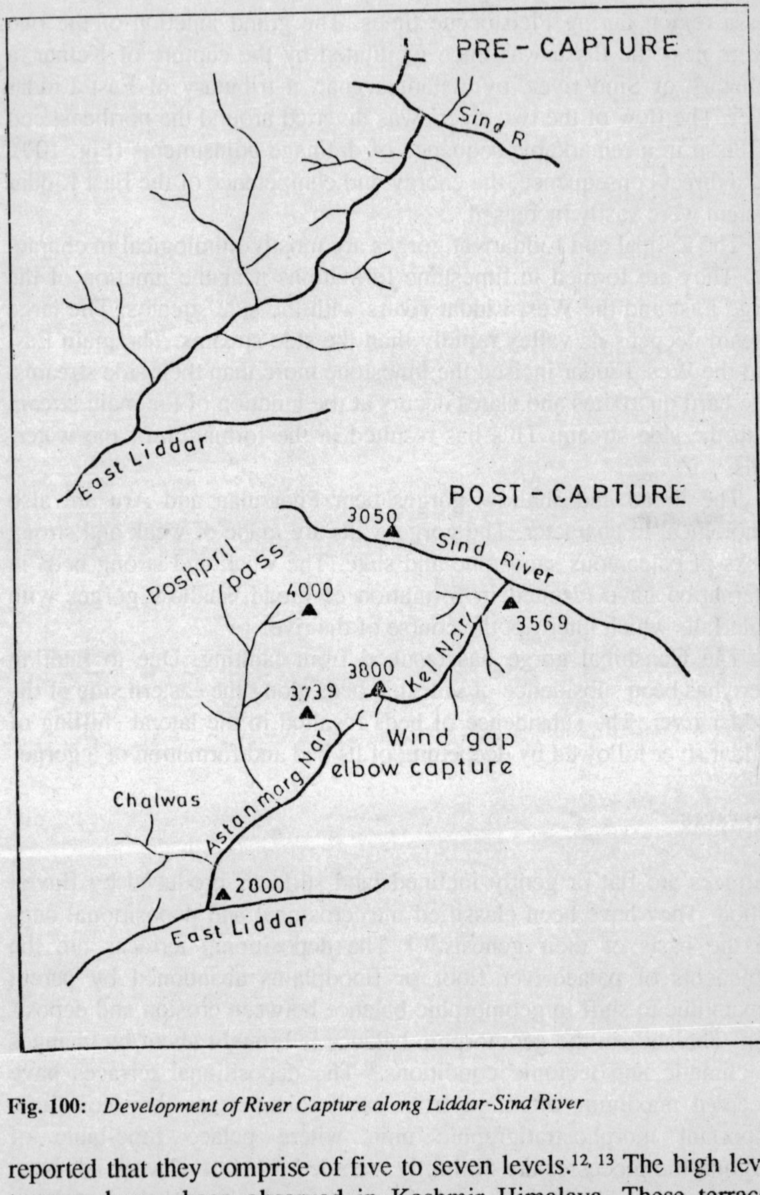

Fig. 100: *Development of River Capture along Liddar-Sind River*

reported that they comprise of five to seven levels.[12, 13] The high level terraces have been observed in Kashmir Himalaya. These terraces stand 300 to 350 metres above the river bed.[14] The systematic study on terrcaes was conducted by Mukerji on *choe* terraces of the Siwalik hills. From the stratigraphic exposures of terraces he reconstructed its palaeoclimatic history.[15] Joshi and Rajaguru were the first to investigate the terraces of the study area.[16] They reconstructed quaternary history of Pahalgam area from the terrace deposit. Terrace fragments are found throughout the whole length of upper and middle reaches of the valley studied. In the middle reach they are best developed in the narrow section from Ganshibal gorge to Aishmaqam. The three terrace levels are identified up to 70 metres above the present Liddar river. They are mostly fluvial in origin (Fig. 96). In the upper reaches of the valley the terrace appears to be glacio-fluvial in origin and the features are produced by the blocking of the main valley by advancing tributary glaciers.

The Liddar terraces in genesis and morphology are similar to the main Himalayan terraces, but are lower and smaller than the Himalayan terraces. The terraces are mostly noncyclic and their downstream slope similar to that of the Liddar river.

The entire morphological history of terraced plains of the Liddar valley clearly points to its evolution being controlled by delicate balance tectonic uplifts and processes of deposition and erosion during the Pleistocene and subrecent periods.

Distribution

Terraces occur in fragments in the East and the West Liddar valleys and are located between the altitude of 2100 and 2800 metres above the mean sea level. Their distribution is highly selective and localised between Pharsulan and Chandanwari along the East Liddar and Mundlun and Aru along the West Liddar river respectively (Fig. 96). The terraces extend in northwest-southeast and northeast-southwest direction. They are glacio-fluvial in origin and they extend from the existing rivers up to the moraines. The terraces are small and narrow. Their downstream slope of about 15° to 20° is similar to those of the East and West Liddar rivers. In the Pahalgam basin two terrace plains are found between the altitude of 2150 metres and 2200 metres near Pahalgam Golf Course and Nunwun.

The terraces are well developed through the entire belt of the middle reach of the Liddar valley. They completely fill the valley and extend right up to the foot of the retreating mountain slopes separating

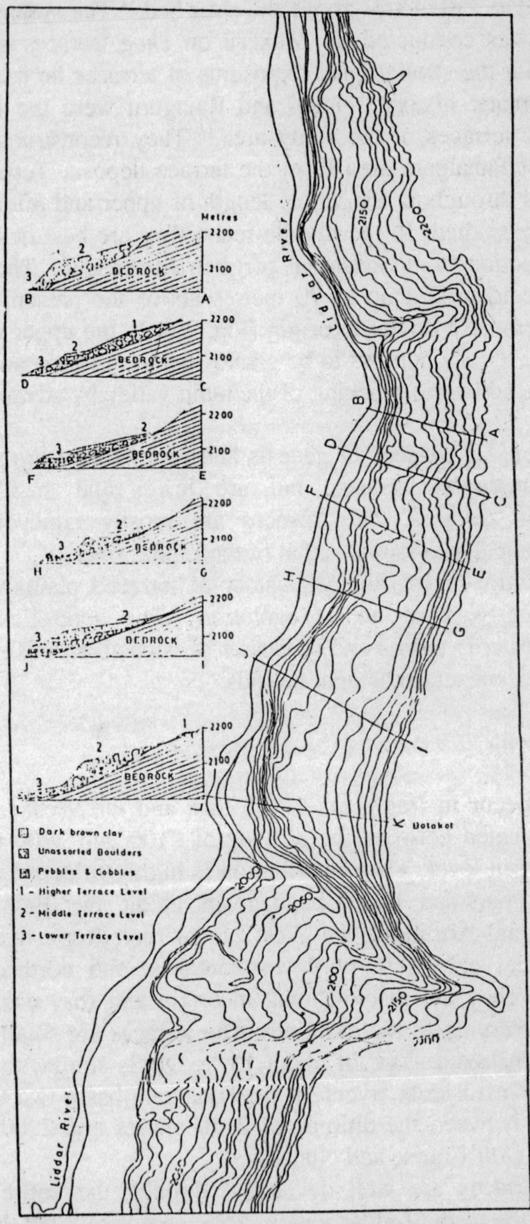

Fig. 101: *Distribution and Morphostratigraphy of the Terraces along the Middle Course of the Liddar River*

them by good distance, from the existing Liddar river. The largest terrace plains are concentrated near the villages of Batakot, Ganeshipur, Salar, Yanar and Aishmaqam (Figs. 96 and 101). In the vicinity of Batakot and Ganeshipur all the three levels are encountered, symmetrically flanking the Liddar river. To the east of the river, the middle terraces form a unilinear strip all along the river course (Fig. 101). The remaining terraces are not continuous and comprise narrow lensoid strips. The low terraces are visible a few metres away from the river and are found in the form of different terraces. The terraces extend in the east-west and west-east directions.

Morphostratigraphy

The morphostratigraphy of the terraces differ from each other and between different sites. The variation are due to changes in the flow pattern of Liddar river that occurred during the Pleistocene times. These

Fig. 102A: *Middle (A) and Lower (B) Terrace Levels Separated by a Low Scarp. The locality is west of Salar*

Fig. 102B: *Small Cuspate Terrace (A) with Middle Terrace Level Surrounding it near Batakot*

changes are recorded in the geometry of the terraces. The width of terraces reflects the lateral shifting of the river in palaeotimes and the height of the terraces scarp documents the vertical erosion. The estimates of the palaeodischarge can also be evaluated from these two components.

Geometry of the Terraces: The geometrical form of the terraces of the middle reaches of the Liddar river are rectangular, semicircular or cuspate, each of which are related to different set of genetic processes. The majority of the lower (T3) and upper terrace levels (T1) have crescentric form. Those of the lower terraces are related to the meander loops of the river. The upper terraces are located near the lower mountain slopes or spurs (Fig. 102). The middle terrace levels (T2) are generally rectangular in form (Fig. 103).

The lower terraces (T3) at Batakot (Hargam), Ganeshipur and Salar are cuspate or semicircular in outline (Fig. 102A). At Batakot the cuspate type terrace is nearly 800 metres long, 80 metres wide and

3 metres high from the river surface. It is mainly composed of gravels and boulders brought from higher reaches of the Liddar. A majority of the lower terraces are 1 to 10 metres high from the existing Liddar river and generally stretch over longer distances. They have a mean width of 135 metres. The lower terrace levels comprise a number of small terraces (Fig. 102B). The terrace surfaces are very irregular and are 2 to 5 metres high from the present Liddar river. They are usually narrow and their terrace scarps are 1/2 to 1 metre high from the river.

The middle terrace (T2) assumes very extensive dimensions in the eastern part of middle stretch, extending uninterruptedly from Batakot in the north to Ganeshipur in the south. In fact this terrace level is most prominent surface occupying the broadest portion of the middle reach of the Liddar valley. The middle terrace levels are mostly rectangular. They are 300 to 600 metres wide, 400 metres to 2 kilometres long and 12 to 30 metres high above the present Liddar river (Fig. 103). The ter

Fig. 103: *Middle Terrace Level Along East of Liddar River Forms Unilinear Rectangular Strip. It is backed by mountain slope*

races are located at comparatively lower altitudes (1925 metres) along the eastern side than in the western side (1970 metres) of the river. The average height of the middle terrace of the eastern side are 18 metres and that of the western side are 26 metres above the present Liddar river. The middle terrace plains are gently inclined and their thalweg gradient is 3° to 5°. The terrace scarps are 3 to 6 metres high from the lower terrace level.

The higher level terraces are asymmetrically distributed between Batakot and Ganeshipur and are mainly localised along the east bank of the Liddar between the altitude of 1970 and 2110 metres. The terraces are nearly 65 metres wide, 200 to 300 metres long and 50 to 70 metres high from the river bed. The terrace surfaces are steeper than the middle terrace and are inclined at 10° to 15° towards down valley.

Stratigraphy: The soil profile of the higher terrace exposed near Ganeshipur terrace scarp has 25 centimetres thick layer of greenish sand and clay at the top followed by 10 centimetres thick pebble sand

Fig. 104A: *Exposed Stratigraphy of Higher Level Terrace Shows Angular Boulders in the Soil Profile*

bed. A large column of boulder beds stretches below the pebble sand bed. The boulders are more than 2 metres diametre and are angular to subangular form. Lenses of green sand are observed near the upper layer of boulder bed (Fig. 104).

The middle terraces have nearly similar soil profile. The lower terrace levels have entirely different soil profiles. The topmost layer of terrace scarp near river bank has 40 centimetres thick clayey-loam layer, followed by 10 centimetres thick pebble-cobble bed. Within this horizon the lenses of sand are seen in the terraces. Below pebble-cobble bed stretches a large column of boulder bed. The boulders are rounded to subrounded in form. The lenses of pebbles and cobbles are concentrated at few sites (Fig. 104B).

The higher and middle terrace levels are generally composed of subangular boulders of slates, limestones and volcanics. The diametre of these boulders are more than 2 metres and are considered to be derived from higher mountain slopes. The shape of the boulder

Fig. 104B: *Exposed Stratigraphy of Lower Level Terrace shows Pebble Lenses (A) in the Stratified Members of rounded Boulders and Cobbles (B)*

indicates that it has travelled a shorter distance or has been transported by flashflood. The boulders are scratched by striae which indicates their glacial origin. But their angular shape reflects that glaciation has not been extensive in the area. The rounded and subrounded form of boulders in the lower terrace levels suggests that these boulders have been formed by gradual erosion before being carried by the river. It also suggests that boulders could have travelled a long distance. But the source area of all boulders is the same locality. Therefore, the distance travelled by boulders is somewhat similar. The subrounded and rounded boulders of the lower terraces could have been moulded by glaciers extensively before being transported by river. It also reflects the extensive nature of the glacier. The boulder beds belong to interglacial phase and thus the formation of terraces began during the interglacial phase of the three glacial advances. The successive layers of pebbles, silty clay and sandy clay suggest moist interglacial and dry glacial periods respectively.

Origin

The distribution and morphostratigraphy of different terrace units of the Liddar suggest their origin in different genetic processes operating during the Pleistocene and Holocene periods. The Liddar terraces are mostly non-paired and are asymmetrically distributed.

The non-paired terraces suggest that the Liddar river has been shifting its flood plain due to rejuvenation that lead to the cutting of the terraces. The rejuvenation could have been caused by continuous uplift of the study region. In Kashmir valley the main outcrops of Karewa is found at an elevation of 2000 metres but one outcrop is found at 3660 metres.[17] This suggests an uplift of Kashmir Himalaya at the rate of 0.1 millimetre per year.[18] The rate of uplift would in any case have appreciably affected the morphology of the river system. The river was meandering laterally and at the same time, because of uplift, it was downcutting. In this condition the river could grade its profile sometimes at a steeper or sometimes at a gentler slope. In either of these cases terraces will not be distributed equally. Their preservation at some places but not at others is also due to the vicissitudes of subsequent erosion.[19]

There are many terrace and terracet levels in the study area but they broadly fall into three main groups located between the altitudes of 1930 and 2200 metres above the sea level. These three levels are identified up to 70 metres above the present Liddar river. All the terraces developed in the middle Liddar river course are fluvial in origin, while those developed in the upper river course are glacio-fluvial in origin. The genesis of the three terrace levels are related to three glacial advance and retreats experienced during the Pleistocene and Holocene period.

De Terra and Wadia have documented four to five glacial advances followed by four retreats in the study region during the Pleistocene period, on the basis of Karewa (lacustrine) terraces.[20, 21] Their findings have been contradicted by recent workers on the basis of sedimentological analysis of Karewa deposits.[22]

The ice age in Liddar valley commenced in the early part of the middle Pleistocene, when the rise of the Himalaya lead to rapid upliftment of Liddar basin. The first glacial advance occurred in the early part of the middle Pleistocene. This phase synchronises with the late Mindel phase of Alpine glaciers.[23] The three glacial phases are therefore related to Mindel, Riss and Wurm and three terraces are related to Mindel-Riss, Riss-Wurm, and Wurm-Holocene interglacial phases. The width of the terraces reflect the moist condition and discharge potentialities of interglacial period. Therefore the width of the Liddar terrace levels suggest that middle terrace level developed during second interglacial (Riss-Wurm) phase was more moist and extensive than the first (Mindel-Riss) and the third (WUrm-Holocene) phase. During the second interglacial phase (Riss-Wurm) the deposition of material was more pronounced towards southwest than southeast as in the first interglacial phase. The shift in deposition was due to lateral swing in the Liddar river from the east to the west. It could be due to further upliftment of the Liddar mountains and development of fault plain in the southern part of the basin. During this phase most of the higher level terraces developed in the first interglacial along the west bank were eroded away due to westward lateral swing of the river. It was followed by the deposition of the middle terrace level on both sides of the river.

The presence of angular to subrounded boulders with final soil fraction of clay and sand at the top of the middle terrace level (T2) suggests an increasing aridity not indicated by the lower (T3) and higher (T1) terraces. The high aridity could have led to extensive growth of glaciers and their maximum advance to Ganshibal during the third glacial (Wurm) advance. During the third interglacial (Wurm-Holocene) the ice retreated slowly in phases. The slow retreat resulted in gradual growth of lower terrace levels and this also favoured the development of terracets on either sides of the Liddar river. During the third inter-glacial phase two glacio-fluvial terraces were deposited in the upper reaches of the Liddar valley.

NOTES

1. Medlicott, H.B. (1868). 'The Alps and Himalayas: A Geological Comparison', *Quarterly Journal of Geological Society of London*, Vol. 24, pp. 34-52.
2. Pal, D. and Srivastava, M.C. (1982). 'Landform configuration of the Karewa Floor and its implication on the Quaternary Sedimentation—Kashmir Himalaya', *Himalaya: Landform and Processes*, Verma, V.K. and Saklani, P.S. (Eds.), New Delhi; Today and Tomorrow Printers and Publishers, pp. 47-48, ref. p. 71.
3. Mithal, R.S. (1968). 'The Physiographical and Structural Evolution of the Himalaya', *Mountain and Rivers of India*, 21st International Geographical Congress, New Delhi, pp. 41-81, ref. pp. 47-51 and 67.
4. Cotton, C.A. (1948). *Landscape: As developed by the processes of normal erosion*, New Zealand; Whitcombe and Tombs Limited, pp. 1-509, ref. pp. 146-154.
5. Lamplugh, G.W. (1908). 'The gorge and basin of the Zambezi below the victoria Falls, Rhodesia', *Geographical Journal*, Vol. 31, pp. 133-152, 287-303.
6. Hansen, W.R. (1965). 'The Black Canyon of the Gunnison', *Bulletin, Geological Survey*, No. 1191, pp. 1-76, ref. p. 15.
7. Russel, R.J. (1940). 'Quaternary History of Louisiana', *Bulletin, Geological Society of America*, Vol. 51, p. 1214.
8. Zeuner, E.F. (1959). *The Pleistocene Period*, London; Hutchinson, pp. 1-447, ref. p. 47.
9. Butzer, K.W. (1976). *Geomorphology from the Earth*, New York; Harper and Row, pp. 1-463, ref. p. 165.
10. Johnson, W.H. (1976). Quaternary Stratigraphy in Illinois: Status current problem', *Quaternary Stratigraphy of North America*, Mahney, W.C. (Ed.), Penn: Stroudsburg, pp. 161-198.
11. Horman, K. (1974). 'The terraces of Seitikhola: Contribution to Quaternary morphogenesis in central Nepal', *Erdkunde*, Vol. 28, No. 3, pp. 161-176.
12. Khan, A.A. and Srivastava, M.C. (1981). 'An evaluation of textural parameters of quaternary terraces, Dhauli Ganga, Garwal, Himalaya, U.P.', *Symposium on Three Decade Development in Petrology, Minerology and Petrochemistry in India.*
13. Vohra, C.P. (1970). 'A note on the terraces of Sutlaj River in Dhar - Slappar area district Bilaspur (H.P.)', *Centre of Advance Study in Geology Publication*, Vol. 7, pp. 119-126.
14. Nawani, P.C. and Khan, A.A. (1982). 'Report on Quaternary geology and geomorphology of Chenab basin between Kishtwar and Ramban, J.K., *Geological Survey of India Publication*, No. 12, pp. 1-23.

15. Mukerji, A.B. (1976). 'Choe terraces of the Chandigarh Siwalik hills', *Revue de Geomorphologie Dynamique*, Vol. 25, No. 1, pp. 1-17.
16. Joshi, R.V. Rajaguru, S.N., Pappu, R.S. and Boparadikar, B.P. (1973). 'Quaternary Glaciation and Palaeolithic sites in the Liddar Valley (Jammu-Kashmir)', *World Archaeology*, Vol. 5, No. 3, pp. 369-379.
17. Jhingran, A.G. (1981). 'Geology of the Himalaya', *The Himalaya: Aspect of Change*, Lall, J.S. and Moddie, A.D. (Eds.) Delhi; Oxford University Press, pp. 77-99, ref. p. 93.
18. Marh, B.S. (1986). *Geomorphology of the Ravi River*, New Delhi; Inter-India Publications, pp. 1-86, ref. p. 9.
19. Johnson, D.W. (1944). 'Problems of terrace correlation', *Bulletin, Geological Society of America*, Vol. 55, pp. 793-818.
20. De Terra, H. and Paterson, T.T. (1939). *Studies on the Ice Age in India and Associated Human Cultures*, Washington, D.C.; Publication Carneigic Institute, No. 493, pp. 1-345.
21. Wadia, D.N. (1941). 'Pleistocene Ice Age deposits of Kashmir', *Proceedings of the National Institute of Science of India*, Vol. 7, pp. 49-60, ref. pp. 55-58.
22. Singh, I.B. (1982). 'Sedimentation patterns in the Karewa basin, Kashmir valley, India and its geological significance', *Journal of Palaeontological Society, India*, Vol. 27, pp. 71-110.
23. See ref. 8, pp. 47-52.

9

Palaeogeographic Reconstruction of the Liddar Valley Glacial Landscape

EARLIER VIEWS ON THE GLACIAL HISTORY

There is some divergence of views on the maximum extent of ice in the Liddar valley during the Pleistocene Epoch and about its glacial history. De Terra and Paterson favoured the extension of ice up to Coeval Karewa lake (5000 sq km), which filled the present basin of the Kashmir valley. They found on the basis of lacustrine deposits, four distinct glacial advances, each followed by retreat.[1] The subsequent field investigations by other scientists indicate, that findings of De Terra and Paterson need to be modified.[2, 3, 4] Gill suggested that Boulder Conglomerate zone adjacent to the Himalayas is much older than the fourth glaciation and no unconformity can be identified in separating the two.[5] Tandon *et al.* state that Karewas are fluvio-lacustrine in origin.[6]

Grinlinton suggested that there were two glacial epochs in the Pleistocene history of the East Liddar valley.[7] The second epoch occurred at a low level. During this epoch there were many advances and retreats. The maximum advance of glacier was up to Pahalgam.[8] Ahmad and Hashmi, on the basis of bench cuts in the valleyhead, proposed three glacial advances of Kolahoi glacier (West Liddar).[9] The last advance was extensive. During this phase the glaciers extended up to Pahalgam. This was followed by slow retreat.[10] According to Joshi there was only one glacial phase in Pahalgam area. The boulder beds of Pahalgam, Krungus and Bawan are of Middle Pleistocene age.[11] Sankalia discovered Early Palaeolithic stone implements from two sites near Pahalgam and ascribed them to the first two interglacial age.[12]

RECONSTRUCTION OF GLACIAL LANDSCAPE

For the reconstruction of the glacial history it is desirable to have uninterrupted stratigraphic data for a long period. The main source of the data is imprinted in the climatic flora zone of the valley, which monitors the fluctuations of the glacier. The glacial history and its fluctuation records are therefore reconstructed from the following evidences:

1. Fossil flora
2. Pollen analysis
3. Lichenometric analysis
4. Glacial geomorphology

Fossil Flora

The fossil flora provides, directly and indirectly, data on environment and glacial fluctuations. The plant macrofossils consist of wood fragments, leaves, seeds and fruits. They usually reflect local environments of derived autochthonous deposits. Mixed assemblages occur in the autochthonous deposits and interpretations vary according to transportation process.[13] The fossil flora of Kashmir valley in general and Liddar valley in particular has been investigated in detail by palaeobotanists.[14, 15, 16] The knowledge of the Pleistocene flora has been derived from the mega fossils of the leaves, the seeds and the fruits, exposed at various altitudes of the valley. From these plant fossils the palaeovegetation has been reconstructed (Fig. 105). The palaeoforests comprised largely of oak, in which *Quercus semecarpifolia* and *Quercus incana* were dominant. The other species of oak prevalent at that time were *Betula utilis* and *Populus nigra*. The conifer constituents consisted of *Pinus excelsa* and *Abies webbiana*. There was a large frequency of *Pinus excelsa*. The forests were mainly comprised of pine-oak species.[17, 18] The conifers constituted the upper canopy, while the lower storey was largely constituted of evergreen and deciduous broad leaved tree species. From these vegetation patterns the following conclusions were arrived at regarding the climatic and orographic changes:

1. There was one single phase of glaciation followed by three interglacial stages.[19] Prior to glaciation there was no pine, but there were oak-laurel woods, which survived during the earlier weak glaciation.
2. The climate at that time was tropical monsoonal because of the lower altitudinal extent of the Himalaya during the Early Pleistocene.

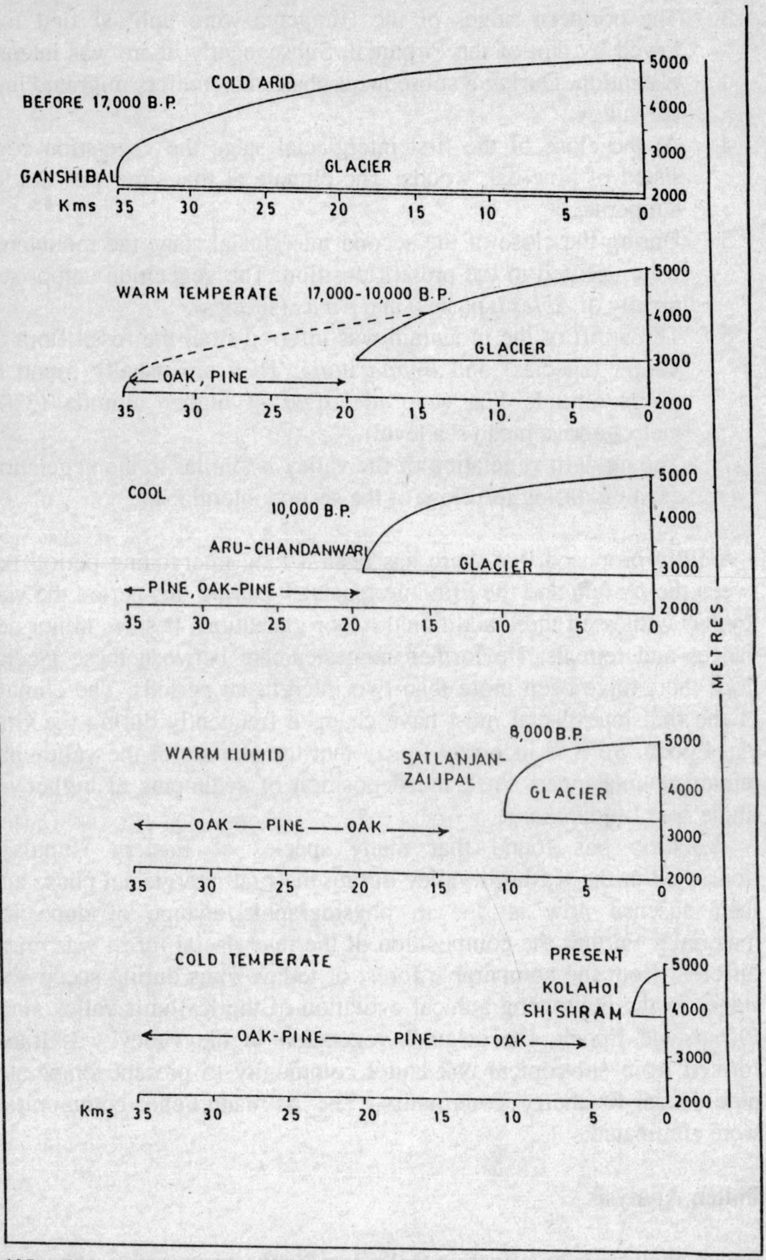

Fig. 105: *Extent of Glacier, Past Climates and Vegetation of Liddar Valley*

3. The northern ridges of the Himalaya were uplifted first followed by that of the Pirpanjal. Subsequently, there was intense glaciation. During a subsequent phase the conifers migrated into the valley.
4. At the close of the first interglacial stage the vegetation consisted of pine-oak woods. The climate at that time was mildly temperate.
5. During the close of the second interglacial stage the mountains were uplifted to the present position. The vegetation comprised mainly of *Abies* (species) and *Pinus* (species).
6. The uplift of the mountain was inferred from the fossil flora of *Alders* (species) and *Betula utilis*. They are usually found at lower altitude but were identified at higher altitude (3500 metres above mean sea level).
7. The modern vegetation of the valley is similar to the vegetation existing during the close of the second interglacial.

Mittre proposed that there has been a long intervening period between the present and the first interglacial.[20] During this period the valley has witnessed three additional major glaciations besides minor advances and retreats. He further suggested that between these glaciations there have been more than two interglacial periods. The climate of the first interglacial must have changed frequently during the Glacial Epoch. So it is incorrect to say that the climate of the valley has remained unchanged since the deposition of sediments at higher altitude near Liddar *marg*.

Kachroo has found that many species of Eastern Himalaya flourished in the Kashmir valley during the first interglacial phase and their absence now is due to physiographic change of deposited material.[21] Further the composition of the interglacial forest was much different from the comparable forest of today. Thus during successive stages in the phytogeographical evolution of the Kashmir valley since Pleistocene Epoch, the original vegetation in the valley was transformed from subtropical oak-laurel community to present temperate-pine-deodar-fir-cherry community. The oak-laurel-chir communities were eliminated.

Pollen Analysis

Pollen analysis is an effective technique for the construction of vegetational history and dating of sediments. Pollens remain preserve for several million years in sediments if protected from oxidation. Also pollen rain related to the vegetation, produced during the flowering of plants, is distributed over relatively large areas. The pollen is deposited in different zones of the stratigraphic units. From the pollen zones it is possible to reconstruct the vegetation map of each stage, which throws light on the prevailing climatic conditions. It further reflects the distribution of ice during a particular stage.

The bog of Liddarwat has been investigated for pollen analysis.[22] The pollen stratigraphy of the region was divided into different vegetation stages on the basis of pollen percentage of different species. This enabled the reconstruction of the vegetation and glacial history of the Liddar (Fig. 106). Since no radiocarbon dating was conducted on Liddarwat bog, the radiocarbon dates of Toshmaidan and Butpathri have been accepted as representative of the region. Their pollen stratigraphy closely correlates with that of the study region. They are located at 50 and 125 kilometres northwest and southwest of the region respectively.[23, 24]

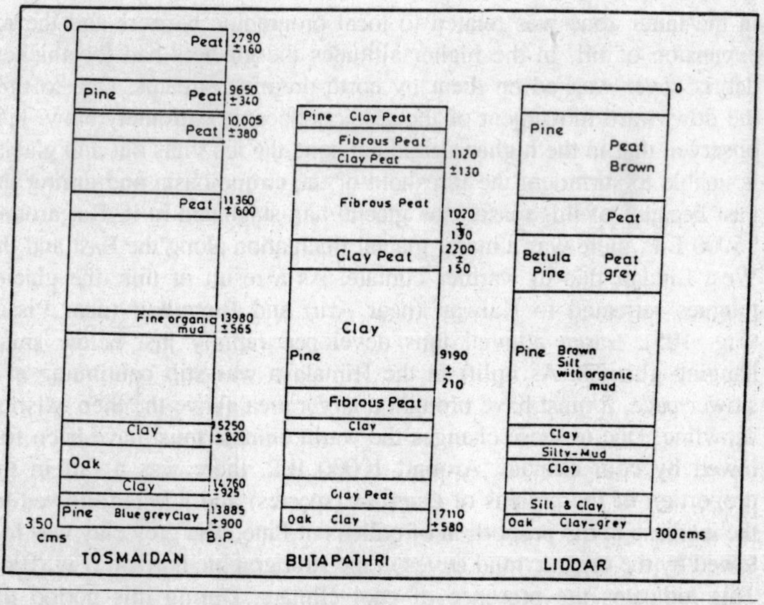

Fig. 106: *The Selected Pollen Diagrams and the Profiles of Sediments of the Kashmir and the Liddar Valley*

The lowest segment of the Liddarwat bog (300-280 centimetres) reveals an extremely high value of *Pinus wallichiana* pollens, but in certain sections (285 centimetres) oak shows upward trend. Oak is represented by *Quercus semecarpifolia*. The overwhelming presence of pine and oak pollens indicate cold arid climate (Fig. 105). The tree constituents did not grow locally near Liddarwat, but occupied, at some distance, the relatively lower slopes near Ganshibal. At Liddarwat the condition should have been quite cold. This stage was followed by sudden fall of pine pollens accompanied by gradual rise of oak pollens. In spite of considerable fall in the proportion of pine pollens, pine still dominated. The overwhelming increase in the proportion of pollens of the broad-leaved trees indicates the presence of warm temperate climate around (17000 B.P.) (Figs. 105 and 106). During this period there was general disintegration of large glaciers which had extended to Ganshibal. Around 17000 B.P. the glaciers which had their snout near Ganshibal retreated to higher altitude. Around 15000 B.P. the glacier near Tuliyan, Chandanwari and Dudhal in the East Liddar valley and Aru and Girwar in the West Liddar valley, retreated from lower altitude zone to the higher altitude zone. The delay in the retreat in the latter zone was related to local orographic barriers and the re-expansion of till. In the higher altitudes the glaciers had the thickest debris cover washed on them by north draining streams. Due to this, the downward movement of the glaciers became extremely slow. It is observed that in the higher altitudinal zone the ice thins out and glacial is unable to surmount the threshold of the cirque basin and during the past because of this reason the glacial had stagnated in it. But around 15000 B.P. there was a major glacial fluctuation along the East and the West Liddar, due to warmer climate. As a result of this, the glacier tongues retreated to Kalwan (near Aru) and Burzulkut (near Pissu) (Fig. 105). Large alluvial fans developed rapidly just below small hanging glaciers. As uplift in the Himalaya was still continuing at a slower pace, it must have brought a larger area above the then existing snowline. Due to these changes the warm climate must have been followed by cool climate. Around 10000 B.P. there was a fall in the proportion of the pollens of *Quercus* (species) and it was followed by the increase in the proportion of pollens of Pine. The grey clay was followed by the organic mud facies in the stratigraphic horizon (Fig. 106). This indicates the presence of cool climate. During this period the glaciers advanced again up to Aru in the West and Chandanwari in the East Liddar, covering a distance of 2 kilometres (Fig. 105). The cool climate was followed by a gradual increase in temperature around 8000 B.P. when the pollen of broad-leaved trees showed marked rise accompanied by a decline of pine pollens to 25 per cent. As a result of this, conifers as a whole, become subordinate. The pollens of *Betula bhojpatra* increased rapidly and attained high value (Figs. 105 and 106). This vegetation indicates warm temperate climate with pronounced rainfall. The heavy rains must have resulted in extensive erosion, followed by removal of soil from alluvial fans in higher altitudes and, further, it must have produced fans at lower altitudes due to freshly exposed deposits. It is because of this reason that the higher altitude fans are of ablation type. The glaciers must have retreated initially slowly to Zaijpal and Satlanjan and then later due to high rainfall retreated quickly to the present position. The higher rainfall might have resulted in high erosion leading to barren pollen profile at 150 centimetres horizon. The top of the pollen profile is marked by complete rejuvenation of conifers with re-establishment of pine pollens[25] (Fig. 106).

Lichenometric Analysis

Lichenometry is the means of estimating the growth of Lichens. The growth helps in determining the age of the depositional feature produced by glacier. The growth rate of lichen is measured from thali that normally develops on rocks. The thallus of lichen comprises groups of fungi and algai, which live together symbiotically in colonies of cells. The cells are held with rock through tubes called hyphae, originating from the fungal spores.[26] Out of different lichen genera only *Rhizocarpon* (species) and *Xanthoria* (species) have been investigated by the earth scientists since they can be easily identified in the field.[27]

Rhizocarpon genus have been investigated intensively as it is most common in alpine environment, and is easily identifiable.[28, 29] The *Rhizocarpon* has variety of species, but frequently they are termed by lichenometrists *Rhizocapon geographicum*. It is yellowish green in colour and its distribution has wide geographic range. It grows at the rate of 1 to 2 centimetres per hundred years but grows to a considerable age.[30] In the initial stages it has faster growth rate (0.15 millimetre per year) up to 300 years, but later its growth retards.[31, 32] Xanthoria is orange in colour, grows well on calcareous substrates. It has a faster growth rate of 2 millimetres per year. The variation in the lichen growth rate responds to variations in amount of moisture, temperature, duration of snow cover and the host rock on which it grows. Lichenometry as a technique was first introduced by Beschel.[33] It is widely used because it can be applied easily and quickly in the determination of dates and needs practically no instrument. However, it is to be applied with a great deal of caution.[34] There is hardly any comprehensive work available on lichenometry of the study area, except the one conducted by the present author in Harmukh mountains, northwest of study region.[35]

Methodology

Lichenometry was used in the study region to reconstruct the regional trend in lichen growth rate and hence develop a framework for dating of moraines and determine the former extension of ice as a function of time. For this study the size of yellow-green *Rhizocarpon geographicum* was measured along the West and the East Liddar in Kolahoi and Shishram mountains above the treeline. Eighty lichen stations were established during the field season and approximately 1000

Fig. 107: *Close-up of* Rhizocarpon Geographicum *Thali (A) in Kolahoi Mountains*

thali were measured (Fig. 107). For each of the station investigated in lichenometric survey, the following measurements were made: elevation, orientation of site, geomorphic feature, rock-type, clast size if the site was comprised of unconsolidated rock, total lichen cover, per cent cover of *Rhizocarpon geographicum* and ten largest thali diameters in the immediate area. The growth curve of *Rhizocarpon geographicum* was developed from three dated localities of Liddarwat, Kolahoi and Shishnag (Fig. 108). Of these, two dated localities are the foundation stones of the Shishnag and Liddarwat resthouses (A.D. 1905, and 1915) and a *Pohal* (shepherd) hut near Kolahoi snout. The foundation stones were cut prior to their emplacement and therefore were lichen free at the time of construction. The estimated growth curve was compared with the growth curve determined in Alps, Colorado Front range and in Alberta.[36, 37, 38] The dated curve of the study region falls between the Colorado Front and Alberta curves, but its growth is somewhat similar to that of Alps (Fig. 108). The mean estimated growth rate of *Rhizocarpon geographicum* is nearly 2 centimetres per hundred

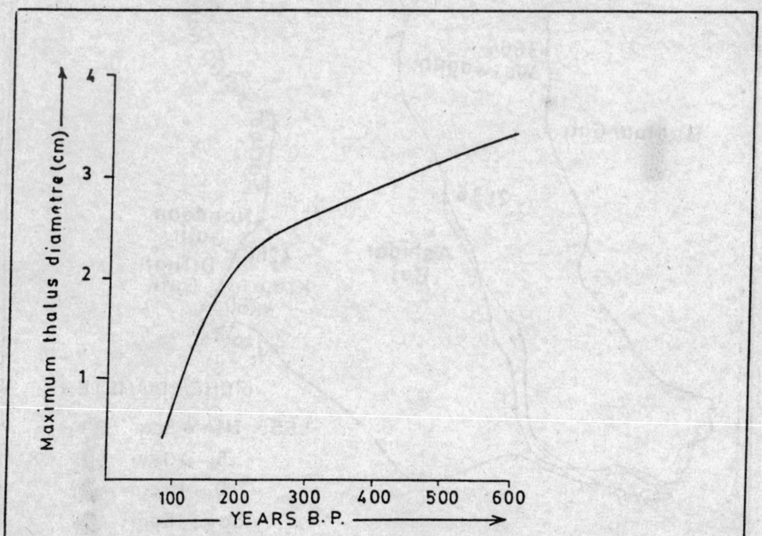

Fig. 108: *Growth Rate Curve for* Rhizocarpon geographicum *in the West Liddar Valley*

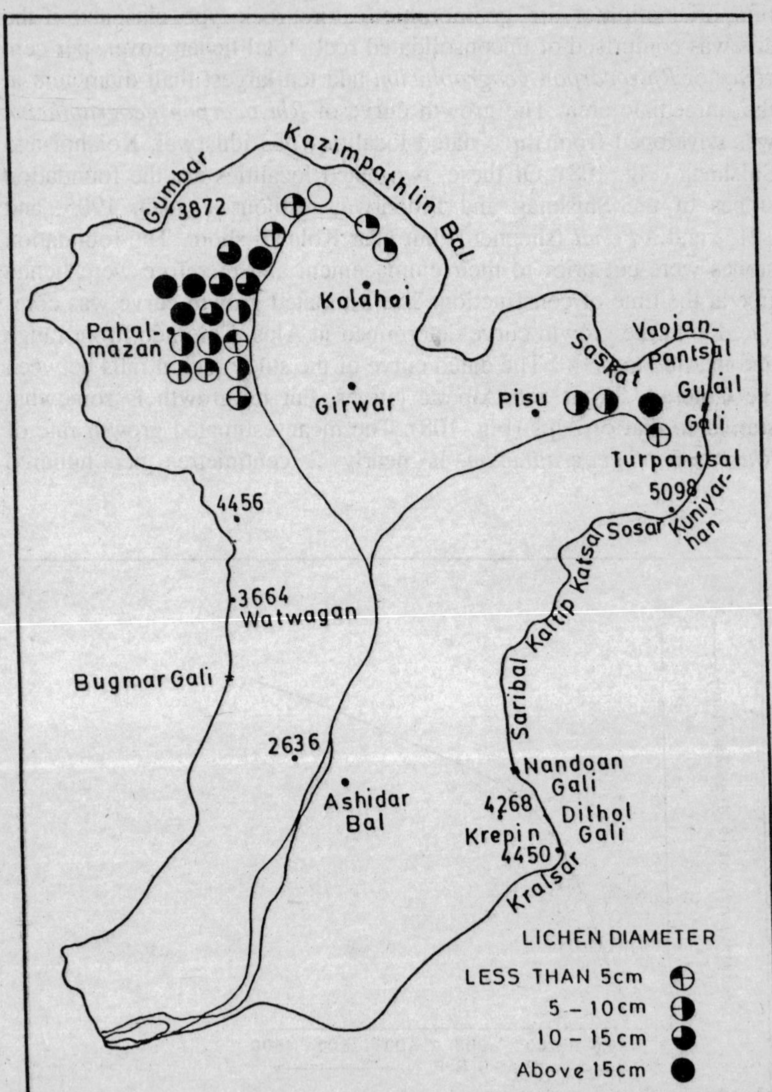

Fig. 109: *Liddar Valley*: Rhizocarpon geographicum *Distribution*

3300 metres, whereas at 3400 metres the low growth of lichens is due to its proximity to the tongue of glacier (Fig. 110). At an elevation of 3500 metres thallus diameter of *Rhizocarpon* is larger than is indicated by its age. The larger growth is related to higher temperature and moisture content during their early growth, but the maximum diameter of old lichens is ecologically controlled by long span of time.

The careful examination of lichen distribution in morainic ridges reveals that the lichens are mainly concentrated in the medium high and low types.[41] In the medium high lateral morainic ridge of the West Liddar the mean lichen diameter ranges from 10 to 13 centimetres and

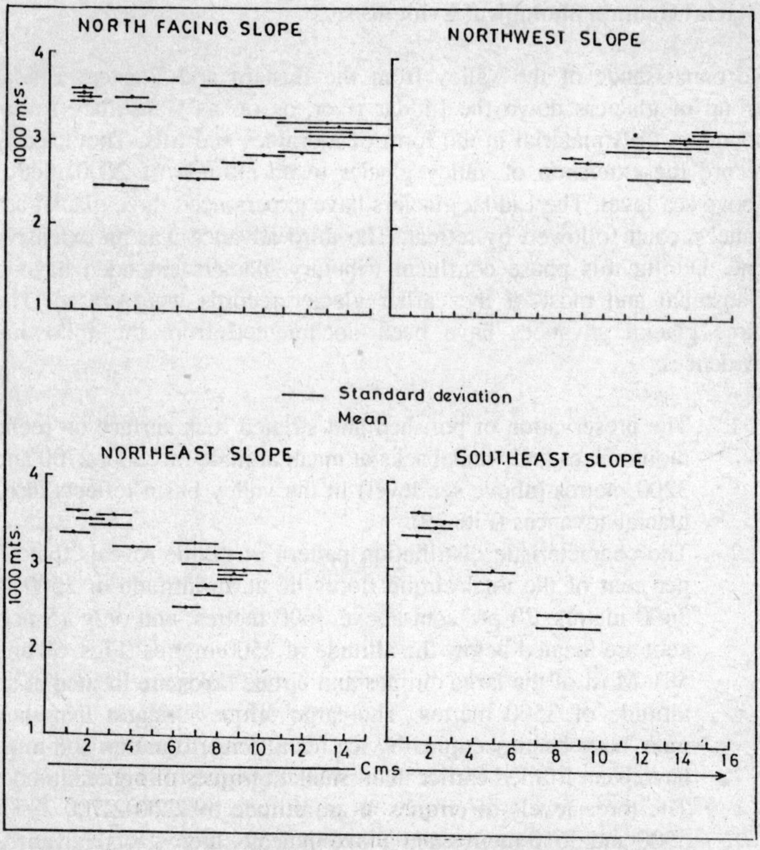

Fig. 110: *Lichenometric Dating of Kolahoi in Different Directions*

their estimated age varies between 6000 and 8000 years. Thus the moraines located near Liddarwat and Satlanjan are nearly 8000 years old. In the East Liddar valley, at Zaijpal and Shishnag, the lichen diameter on morainic boulders ranges between 12 and 13 and 8 and 10 centimetres respectively. Their estimated age ranges from 7000 to 8000 and 5000 to 6000 years respectively.

There is a multiplicity of small, recently formed moraines near the snout of Kolahoi. They are mostly devoid of lichens. About half a kilometre away from the snout there are three low lateral morainic mounds. Its rock blocks are patterned with lichens. Their estimated age ranges between 250 and 500 years.

Glacial Geomorphological Evidences

Reconnaissance of the valley from the Eastern and Western Liddar group of glaciers down the Liddar river, as far as Ganshibal, reveal abundant drift material in the form of moraines and tills. They clearly record the extension of valley glacier to an altitude of 2000 metres above sea level. The Liddar glaciers have experienced three glacial advances, each followed by retreat. The third advance was an extensive one. During this phase confluent tributary glaciers extended beyond Ganshibal and most of the earlier glacier records were erased. The three glacial advances have been documented from the following evidences:

1. The preservation of polished and striated rock surface on roche moutonnees and whalebacks at mean altitude of 2200, 2700 and 3200 metres (above sea level) in the valley basin reflects three glacial advances (Fig. 39).
2. The characteristic distribution pattern of cirque reveals that 65 per cent of the total cirque floors lie at an altitude of 2500 to 3600 metres, 20 per cent above 3600 metres, and only 15 percent are located below the altitude of 2500 metres (Figs. 57 and 58). Most of the large cirques and cirque lakes are located at an altitude of 3500 metres. The large cirques suggest that they must have been occupied by ice for a longer duration and must have been formed earlier than smaller cirques of higher altitude. The three levels of cirques at an altitude of 2200-2700, 2800-3500 and 3650 metres and above indicate three glacial advances (Fig. 58). The cirque levels are synonymous with bench cuts

and glacial scoops observed on valley walls by earlier investigators.[42, 43]

3. Two high lateral morainic ridges of Zaijpal and Liddarwat basin are suggestive of their origin to two glacial advances. The outer ridge is older and the inner one is younger (Figs. 73 and 77).

The inner morainic ridge has been formed during the third glaciation while the outer older one was formed during the earlier two advances.

GLACIAL HISTORY

The glacial sequence in Liddar valley has been divided into stades and interstades. During a stade deep valleys and troughs of the northern part of Liddar valley were occupied by tongues of ice. An interested on the contrary was the time when such glaciers retreated. The main valleys were then occupied by lakes or streams. The three main stades are designated after local places:

1. Ganshibal (Lower Ganshibal)
2. Chandanwari - Kalwan (Middle Ganshibal)
3. Zaijpal - Satlanjan (Upper Ganshibal)

Many kind of evidences have been used in differentiating the stades. Some of the more important ones are: (1) moraines of each of the several stages were less extensive than those of the immediately preceding ones. The moraines nearest the head are the youngest and those farthest are commonly the oldest, (2) the moraines of the youngest stage (Zaijpal-Satlanjan) were generally small in size as compared to bulky embankment of Ganshibal stade, (3) the moraines of the youngest stade were dated with the help of lichenometric data. The moraines near the snout are the youngest (250 years) and those farthest are the oldest. The Zaijpal-Satlanjan stade moraines are 6000-8000 years old. Each of the three stades and interstades are dated by pollen dated techniques. The Ganshibal moraines date to over 17000 B.P. and Zaijpal-Satlanjan moraine to only 8000 years B.P., and (4) the extent to which the terminal moraine has been preserved is also a good evidence. The Chandanwari-Kalwan terminal moraines have not been destroyed by Liddar river as moraines stand as high as 50 metres from the channel bed (Figs. 73 and 74). The Ganshibal moraines, on the other hand, had been breached widely by the river.

Ganshibal Stade (Lower Ganshibal Stade)

The Ganshibal stade synchronises with the third Liddar glacier advance. The areal extension of the various till deposits indicate that Ganshibal stade was the most widespread of all glaciations. It covered an area of 450 square kilometres (Fig. 111). The extension of ice took place much before 18000 B.P., when the climate was cold arid. During this stade Kolahoi group (West Liddar Group) of glaciers extended up to the upper lip of Pahalgam cirque near the race course site. The glacier extended in the north from the Kolahoi at an altitude of 5000 metres, and later took a turn of 260° to move along the West Liddar valley. While on its way to Pahalgam twelve hanging glaciers joined it (Fig. 111). Eight hanging glaciers joined between the snout and Liddarwat and another four between Liddarwat and Aru. Aru glacier was fed by the same *nevz* field as the one which now feeds the Kolahoi glacier. The West Liddar glacier was nearly 150 metres thick. The Shishram group (East Liddar group) of glaciers commenced from Shishram glacier in northeast and later turned 150° to move along the East Liddar valley. The glacier was joined by eight hanging glaciers near Shishnag trough, four near Zaijpal, four near Chandanwari and two near Pahalgam. The East Liddar glacier group was more massive and there ice had a thickness of nearly 200 metres. It extended up to the Pahalgam cirque floor and coalesced with the West Liddar glaciers at an altitude of 2100 metres. The third group of glaciers joined the East Liddar glaciers at an altitude of 2100 metres above sea level near Pahalgam church. It originated from Tuliyan glacier in the east and turned 120° to terminate at Pahalgam and Ganshibal. The glaciers advanced to Pahalgam much later than the Shishram group of glaciers. One branch of the Tuliyan glacier overran the East Liddar glacier near Pahalgam tourist hut complex and the other branch terminated near Ganshibal (Fig. 111). Most of the tributary glaciers extended 4 to 8 kilometres beyond the confluence of tributary valleys. The extension is related to high steep valley floor carved out by glacial erosion, such as the Dudhal, Wokhbal and Tarsar. Their floor gradient ranges between 500 and 400 metres per kilometre. The widely expressed argument that the Pleistocene snowline depression of more than 1000 metres would not appreciably enlarge the glacier to extend beyond the tributary valley is, therefore, difficult to sustain for the study region. There can be no doubt that the substantial valley ice was generated by the growth of large transfluent glaciers.

Interstade

The East and West Liddar group of glaciers retreated around 17000

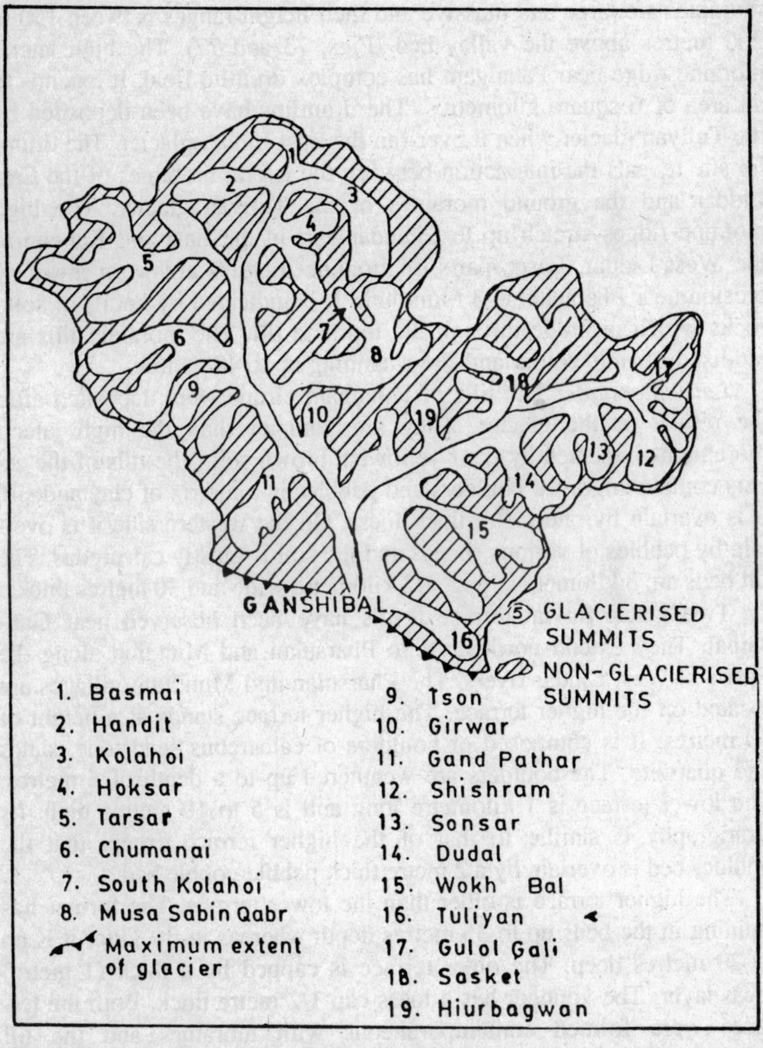

Fig. 111: *Glacier Extent Pahalgam-Ganshibal Stade*

B.P., when the climate ameliorated and the glaciers shrank. It was delayed in the Tuliyan group of glaciers to around 15000 B.P. due to re-expansion of the till of the East Liddar.

Early Interstade: During the early interstade the lateral moraine ridges were formed. They are still fairly well preserved as a bold embankment along the East and the West Liddar river. Near Pahalgam the moraines are large and massive and their height ranges between 150 to 200 metres above the valley bed (Figs. 73 and 77). The high lateral moraine ridge near Pahalgam has complex drumlin field. It extends to an area of 6 square kilometres. The drumlins have been deposited by the Tuliyan glacier when it over-ran the East Liddar glacier. The drumlin site reveals the interaction between the lateral moraines of the East Liddar and the ground moraines of the Tuliyan glacier. The high moraine ridges stretch up to Chandanwari in the East and Kalwan in the West Liddar. Large parts of moraine tongues are worn away by erosion near Pharsulan and Mundlum. It is indicated by knolls of solid rocks which protrude through the moraine till. The morainic tills are reddish brown in colour and show staining up to 40 metres.

Late interstade: The tills of Pahalgam cirque were deposited after the retreat of the glacier. They are younger than the high lateral moraine as indicated by their yellowish brown soil. The tills of the eskers contain engraved boulders and pebbles in a matrix of clay and silt. It is overlain by one metre thick loess. On the western side it is overlain by pebbles of various shapes and the soil is mostly calcareous. The till beds are 3 kilometres long, 1/2 kilometre wide and 30 metres thick.

Two glacio-fluvial terrace levels have been observed near Ganshibal. They extend northwards to Pharsulan and Mundlun along the East and West Liddar rivers. The Pharsulan and Mundlum villages are located on the higher terrace. The higher terrace stands at a height of 40 metres. It is composed of boulders of calcareous sandstone, slates and quartzite. The boulders are weathered up to a depth of 5 metres. The lower terrace is 1 kilometre long and is 5 to 10 metres high. Its stratigraphy is similar to that of the higher terrace except that the boulder bed is overlain by a 2 metre thick pebble-cobble bed.

The higher terrace is older than the lower terrace. The former has staining in the beds up to 35 metres depth whereas in the latter it is up to 20 metres deep. The older terrace is capped by a thick (1 metre) loess layer. The younger has a loess cap 1/2 metre thick. Both the terraces were formed contemporaneous with moraines and the till deposits in the cirque basin. The filled up cirque basin lies inside of high moraine. The oldest terrace corresponds to the belt of high moraines of Ganshibal.

Chandanwari-Kalwan Stade (Middle Ganshibal Stade)

It is the second glacial stade of the valley. During this phase glaciers extended to Chandanwari and Kalwan at an altitude of 2800 to 2400

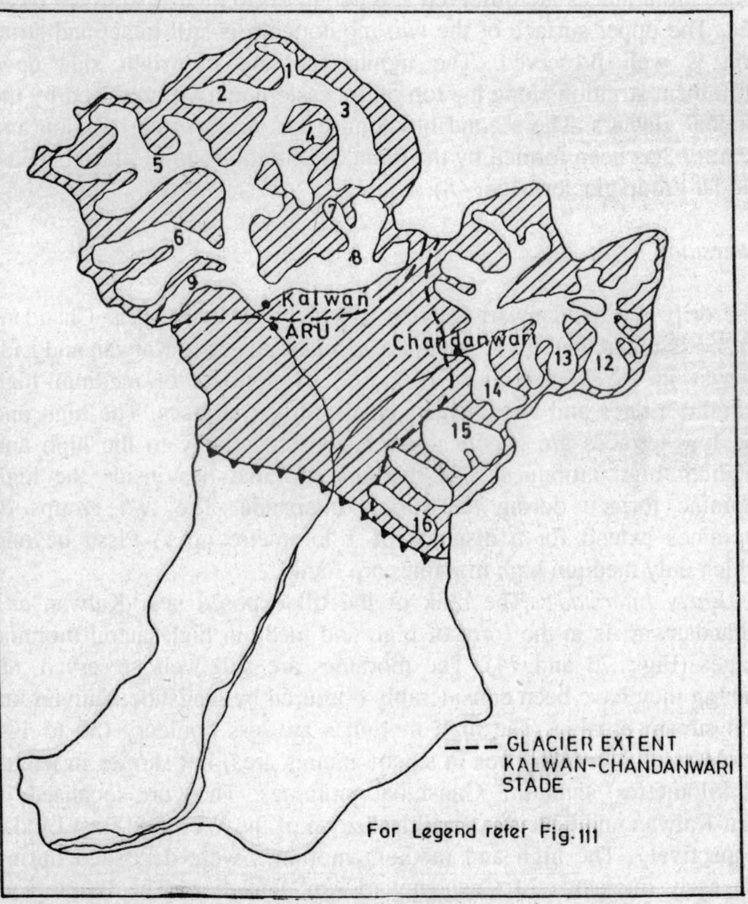

Fig. 112: *Glacier Extent. Kalwan-Chandanwari Stade*

metres (above mean sea level). During this stade the East Liddar group was made of fifteen tributary glaciers up to Chandanwari (Fig. 112). In the Chandanwari basin there was a confluence of two hanging glaciers from east and one from west near two roche moutonnees. The West Liddar group comprised of ten tributary valley glaciers up to Kalwan (Fig. 112). Near Kalwan two hanging glaciers from east and two from west converged near the whalebacks. The cross striation has been observed on the whalebacks as well as on the valley walls of the East and West Liddar at an altitude of 2600 metres. These clearly reflect the changes in glacial phase of Liddar. In the Chandanwari basin two roche moutonnees are observed at a height of 50 metres from the basin bed. The upper surface of the two moutonnees is still intact and striation is well preserved. The moutonnees on the right side have prominent striation along the top of the eastern margin produced by the Dudhal glaciers. The second moutonnee has north-south striation and striation has been formed by the main Shishram group of glaciers (East Liddar group glacier) (Fig. 48).

Interstade

The drift of Chandanwari-Kalwan is widely preserved near Chandanwari, Pissu and Burzulkut in the East Liddar and Aru, Kalwan and Liddarwat in the West Liddar (Fig. 73). It consists of medium high moraine ridges and two distinct glacio-fluvial terraces. The high and the low terraces are genetically related respectively to the high and medium high moraines. The medium moraine lies inside the high moraine formed during the earlier interstade. The two groups of moraines extend for a distance of 1 kilometre up to Pissu beyond which only medium high moraines are found.

Early Interstade: The bulk of the till exposed near Kalwan and Chandanwari is in the form of high and medium high lateral moraine ridges (Figs. 73 and 74). The moraines are still well preserved, although they have been considerably obscured by landslide, alluvial fan and stream erosion. The high moraines are less bouldery (50 to 100 boulders of 30 centimetres in square metres area) but shorter in length (2 kilometres) than the Ganshibal moraines. They are localised in Aru-Kalwan and Chandanwari-Pissu zone of the West and East Liddar respectively. The high and medium moraines were deposited during the early interstade of Chandanwari-Aru glaciation. The lithological and topographical contrast in the two moraines reflect their origin in two confluent tributary glaciers. The medium high moraines of Aru contain the boulders of volcanic trap and calcareous sandstone in a matrix of sand and clay. The clast are sub-rounded and their sphericity is nearly 0.4[44] The high moraines on other hand contain the boulders of sandstone and volcanics. The clasts are rounded and their spericity is nearly 0.6. The lithological characteristics of the high moraines are similar to those of Girwar, Nafron and Hatnar moraines of tributary glaciers. High moraines are steeper (30°-40°) than the medium high moraines (20°- 30°). The lithological and topographical contrasts suggest that the high moraine has its source in Kolahoi, because it is primarily composed of volcanic rocks. The roundedness of the clast is due to its prolonged abrasion. In medium high moraines the source of morainic load is in the neighbouring Musa Sabin Qabar glacier.

The high lateral moraines of Chandanwari contain the boulders of volcanic trap and limestone. The clasts are highly polished and striated. Their sphericity is nearly 0.7. The boulders of the medium moraine are mostly of sandstone and slate. The clasts are subrounded. These moraines are actually continuation of Chalwas and Dudwal tributary moraines. The variations in the two belts of moraine indicate that the high moraines have been formed at the confluence of a tributary glacier within the main Aru-Chandanwari valley glacier. The substantial height of the moraine is due to heavy drift carried by larger transfluent glaciers. In the high and medium high lateral moraines the weathering pits are very large. They cover 70 per cent of the area near Chandanwari and Aru. There is incipient growth of calcium carbonate beneath the boulders. Oxidation is evident below a depth of 20 metres.

Late interstade: The bulk of the medium high lateral moraine ridges preserved in the Barzulkut-Zaijpal and Minimarg-Liddarwat sections of the main valley and at Chalwas, Dudhal, Girwar, Nafron and Tarsar in the tributary valleys were formed in the late interstade of Kalwan-Chandanwari stade (Figs. 73 and 79). The morainic ridges stretch over an area of about 100 square kilometres. They have maximum extension of 6 kilometres and stand at a height of 75-125 metres from the valley bed. The moraines are very bouldery. The frequency of boulders of above 30 centimetres diameter is nearly 150 in an area of 125 square metres. The boulder pits are small and they cover nearly 50 per cent of the area near Zaijpal. Towards Zaijpal the number of pits per square kilometre is small but the diameter of the individual pit is large. This is due to the presence of limestone. The morainic tills show staining up to a depth of 15 metres. In the tributary glaciers such as

Dudhal, Chalwas and Nafron the morainic ridges extend to considerable distance from the valley bed. It is possible to trace them to the cirque where the glaciers originated. The moraines were deposited within these valleys after the formation of the cirques. The Dudhal, Nafron and Chalwas streams have cut through frontal moraines to a depth of 30 to 10 metres.

There was general shrinkage and thining of Liddar glaciers in the late interstadial phase. Most of the small hanging glaciers retreated to their valley heads. The retreat was slow on the protected slopes of East Liddar in comparison to West Liddar valley. In the West Liddar valley most of the hanging glaciers on the western side retreated to higher altitudes leaving behind the cirque basin. The confluent Kolahoi glacier retreated above the Liddarwat and vacated a number of cirques near Minimarg and huge cirque near Liddarwat rest house. In the East Liddar valley the small hanging glaciers retreated in their cirque basins. In Shishnag basin, out of 8 hanging glaciers four tributary glaciers of Gratbal retreated in their cirque basin at an altitude of 3450 metres. They deposited the massive drift along the western flanks of the Shishnag cirque basin at the upper end of the lake. The outlet stream incised the western morainic dam. In Wokhbal, Dudal, Tuliyan and Sonsar the rock-bound mountain cirques were excavated by bare ice.

Terrace fragments have been observed throughout this section, but one pair of two terraces is located near the narrow section of Chandanwari and Aru at an altitude of 2500 and 2700 metres (Fig. 96). Their height varies between 30 and 60 metres and 20 and 40 metres. The high and the low terraces are genetically related to the high and medium high morainic ridges respectively. The bore hole logging of the two terraces reveal that the higher terrace has two horizons of boulder beds separated by a 3 metre thick bed of pebble cobble sledge. The terrace has been drilled up to the depth of 50 metres and no rock bed has been encountered. In the lower terrace only one boulder bed has been found. The rock bed has been encountered at a depth of 25 metres. The terraces are of fill type and consist of glacial drift.

Zaijpal-Satlanjan Stade (Upper Ganshibal Stade)

The well preserved terminal moraine loops, recessional moraine, and whalebacks near Zaijpal and Satlanjan basin at an altitude of 3200 metres (above mean sea level) reflect the recessional halt and minor advance of the Liddar glaciers. The movements took place around

9000 B.P., when there was a sudden change in the climate from warm to cool.[45] The glaciers advanced by nearly 1/2 kilometre from the lip of the waterfall, at an altitude of 3350 metres, to the basin floor at an altitude of 3200 metres. The cross striation on ten whalebacks near the Zaijpal waterfall and three in Satlanjan indicate confluence of tributary glaciers. During the Zaijpal-Satlanjan stade the Liddar glaciers covered an area of nearly 200 square kilometres (Fig. 113). The Kolahoi, the larger transfluent of west Liddar glaciers, extended up to the tip of Satlanjan basin. It had seven tributary glaciers, four confluenced it in the upper catchment area near the Kolahoi Ganj and three near Satlanjan basin. The three hanging glaciers of Satlanjan terminated at an altitude of 3260 and 3300 metres (above mean sea level). They dropped the material over a large area. The terminal moraine is well preserved along the western margin of the West Liddar river between Sosin nar and Ruiyal nar. It has a length of 1 kilometre, a width of 400 metres and a height of 10 metres above the river bed. In the East Liddar valley the transfluent Shishram glacier extended up to Zaijpal at an altitude of 3250 metres. It had nine hanging glaciers — six having confluences in the upper catchment area and three in Zaijpal basin (Fig. 113). Of the latter Sonsar and Natsar joined it in east and Sasket coalesced with its western flank at an altitude of 3250 and 3300 metres (above the mean sea level) respectively. The tributary glaciers dropped the material at their terminal end. It is preserved in the form of terminal loops in the Zaijpal basin (Figs. 73 and 74).

Early Interstade: In the early interstade phase the recessional moraines were deposited along the inner medium high lateral moraine (Fig. 90). These moraines were deposited along the western flanks of Zaijpal and Satlanjan basins. In Zaijpal basin three recessional morain ridges were deposited in a stepped form. The recessional moraines range in length between 1/2 and 1 kilometre and in height between 10 and 25 metres. At the top of the recessional moraine there is a small drumlin field. It stand adjacent to the medium high lateral moraine ridge of Zaijpal. The drumlin field has been formed due to coalescence of the Sonsar hanging glacier with the main tongue of the Shishram glacier. The Shishram glacier retreated earlier and deposited the moraine. Over the moraine the Sonsar glacier deposited the drumlins while retreating to its snout. In the Satlanjan basin two recessional moraine ridges have been deposited during the interstade phase. The moraines are 2 kilometres long and are 5 to 10 metres high.

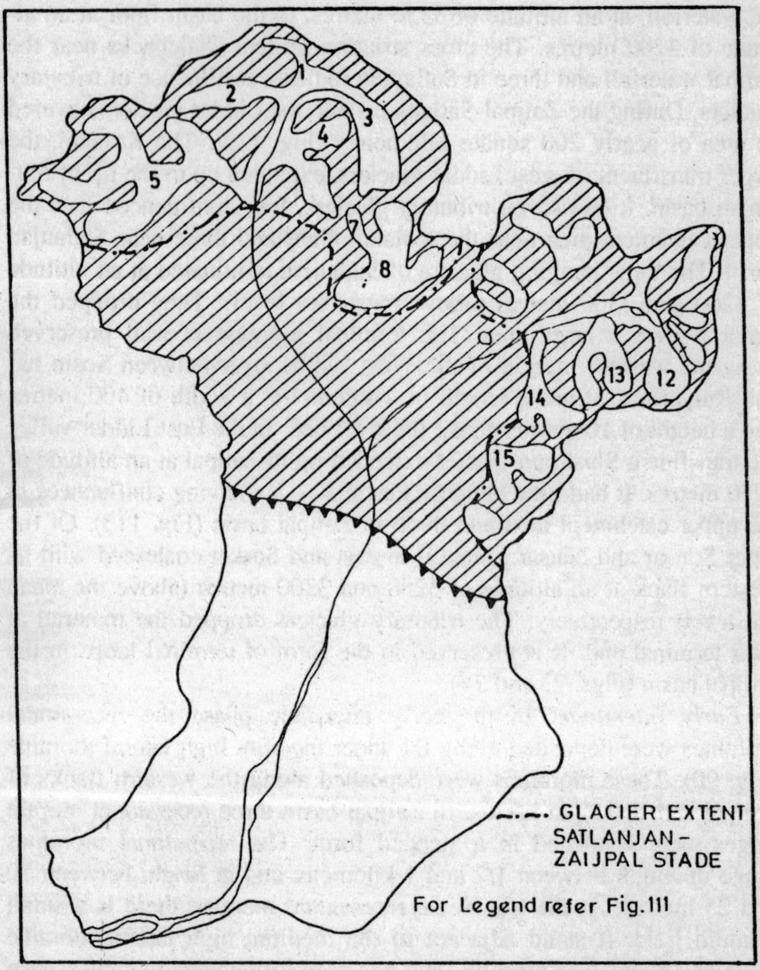

Fig. 113: *Glacier Extent: Satlanjan-Zaijpal Stade*

The recessional moraines are capped by loess to the north of Satlanjan whereas to the south they are breached by deflated alluvial cone. The deflated material is mainly composed of angular blocks of basalt and andesite.

The medium high moraine ridges are located adjacent to the recessional moraines. They are nearly 75 metres high above the valley bed. The moraines are very bouldery. They have steep convex slope towards valley. The slope ranges between 40° and 50°. The large kettle holes have been observed in Zaijpal moraines. The number of kettle hole increases towards Zaijpal waterfall. The largest kettle hole has been observed at an altitude of 3220 metres to the northeast of Zaijpal. It has a diameter of 6 metres. In Satlanjan, the moraines are heavily weathered and are very bouldery. The frequency of angular boulders, with diameter of more than 30 centimetres increased to 200 in an area of 125 square metres. The boulders are severely weathered. The large veins of feldspar protrude out from the disintegrated greywacke boulders. During the early interstadial age and glaciers retreated very slowly and most of the cirque lakes vacated by ice in the earlier stade were nearly filled up. The Minimarg, the Tarsar, and the Liddarwat *marghs* were formed by the deposition of glacial debris. In the Liddarwat, the huge cirque lake carved out in the earlier stade was partly filled by the ground moraine of Tarsar glacier and lateral moraine of Kolahoi glacier.

In the East Liddar valley, the two lateral morainic loops were built on the west side the Shishnag lake basin by the two tributary glaciers of Vaojan and four tributary glaciers of Gratbal. The west side of the Shishnag lake was partly filled up by the lateral moraine of Vaojan-Gulailgali glacier and by the ground moraine of Gratbal glacier.

Middle Interstade: During the middle interstade phase the low lateral morainic ridges were formed. They are well preserved between Zaijpal and Shishnag section of the East Liddar and in Satlanjan-Kolahoi section of the West Liddar. The moraines are mainly bouldery and contain little matrix of ground mass. They are very fresh and are little weathered. The preponderance of boulders on the surface is probably due to excessive erosion by rain. Also, little wind blown loess has been deposited in post-glacial times in the higher altitudes. The low morainic ridges were formed around 8000 B.P. when there was amelioration in climate from cool to warm wet climate. During this time there was heavy rainfall and most of the glaciers retreated rapidly to their cirque basin in their valleyhead. In the West Liddar valley most of the glaciers melted completely on the western side as they were directly exposed to sun rays and, hence, completely melted. In the protected north and northeastern trenches, the glaciers retreated slowly to higher altitude. The Kolahoi retreated to higher reaches of Satlanjan basin.

The Shishram glacier rapidly retreated near the upper reaches of Shishnag lake at an altitude of 3550 metres. Due to quick retreat all the marshy cirque lakes were completely filled up and developed into local *marghs* exemplified by Liddarwat, Tarsar, Basmai, Satlanjan in the West Liddar and Zaijpal, Sonsar, Natsar, and Shishnag in the East Liddar. The Shishnag lake was dammed along the western margin due to further growth of morainic loops. The level of lake rose to maximum height, due to retreating glacier. Its waves caused vertical deepening of the spur line at the exist of the Shishnag lake. Later a V-shaped notch was carved through which the lake water gushed out. At present there are only rock bars.

During this interstade the tills were deposited along the mountain front. They were deflated due to severe erosion by heavy rains.

Late Interstade: In the late interstade stage the small tiny moraines were deposited in the bottom of the valleys along retreating glaciers. These moraines stretch hardly 200 metres in length and stand 5 metres above the valley front. They lie inside the low moraine ridges.

NOTES

1. De Terra, H. and Paterson, T.T. (1939). *Studies on the Ice Age in India and Associated Human Culture*, Washington. D.C.; Publication Carnegie Institute No. 493, pp. 1-345.
2. Singh, I.B. (1982). 'Sedimentation pattern in the Karewa basin, Kashmir Valley, India and its geological significance', *Journal of Palaeontological Society, India*, Vol. 27, pp. 71-110.
3. Ahmad, N. and Hashimi, N.H. (1974). 'Glacial history of Kolahoi Glacier, Kashmir, India', *Journal of Glaciology*, Vol. 13, No. 68, pp. 279-282, ref. p. 279.
4. Bhat, D.K. (1975). 'On the Quaternary Geology of the Kashmir Valley with special reference to stratigraphy and sedimentation', *Geological Survey of India, Miscellaneous Publication*, No. 24, pp. 188-204.
5. Gill, W.D. (1952). 'The stratigraphy of Siwalik series in northern Potwar, Punjab Pakistan', *Quarterly Journal of Geological Society*, London, Vol. 107, pp. 375-394, ref. pp. 381-384.
6. Tandon, S.K. (1981). 'Sedimentology of the Karewa Group of Kashmir Valley : retrospect and prospect', *Man and Environment*, Vol. 5, pp. 1-5.
7. Grinlinton, J.L. (1928). 'Former glaciation of the East Liddar Valley, Kashmir', *Memoir, Geological Survey of India*, Vol. 49, Part 2, pp. 289-388, ref. pp. 379-381.
8. See ref. 7, p. 380.
9. Ahmad, N. and Mayewski, P.A. (1981). 'Himalayan Glaciation : An unresolved issue', *The Geographer*, Vol. 28, No. 2, pp. 21-32, ref. p. 27.

10. See ref. 3, p. 382.

11. Joshi, R.V. Rajaguru, S.N., Pappu, R.S. and Bopardikar, B.P. (1973). 'Quaternary glaciation and palaeolithic sites in the Liddar Valley (Jammu - Kashmir)', *World Archaeology*, Vol. 5, No. 3, pp. 369-379.

12. Sankalia, H.D. (1971). 'New evidence of Early Man in Kashmir', *Current Anthropology*, Vol. 12, No. 45, pp. 558-562.

13. Goldthwait, R.P. (1966). 'Evidences from Alaskan glaciers of major climatic changes', *World Climate from 800 to 0 B.C.*, Sawyer, J.S. (Ed.), London; Royal Meteorological Society, pp. 40-53, ref. pp. 40-41.

14. Puri, G.S. (1984). 'A Preliminary notes on the Pleistocene floras of Karewa formation of Kashmir', *Journal of the Geological Mining and Metallurgical Society of India*, Vol. 50, pp. 61-66.

15. Singh, G. (1963). 'A Preliminary Survey of Postglacial vegetational History of Kashmir Valley', *The Palaeobotanist*, Vol. 12, No. 1, pp. 73-108.

16. Vishnu Mittre and Sharma, B.D. (1966). 'Studies of post glacial vegetational history from the Kashmir Valley', *The Palaeobotanist*, Vol. 15, No. 1 and 2, pp. 200-204, ref. p. 203.

17. Vishnu Mittre (1967). 'Floristic and ecological reconsiderations of the Pleistocene plant impressions from Kashmir', *The Palaeobotanist*, Vol. 16, Part 3 and 4, pp. 308-329.

18. Puri, G.S. (1957). 'Preliminary Observations on Phyto-Geographical changes in the Kashmir valley during Pleistocene', *The Palaeobotanist*, Vol. 6, Part 1, pp. 1-17.

19. See ref. 18, pp. 4-7.

20. See ref. 17, pp. 324-325.

21. Kachroo, P. (1985). 'A note on flora of Kashmir', *Geography of Jammu and Kashmir*, Husain, M. (Ed.), New Delhi; Ariana Publishing House, pp. 32-44, ref. p. 33.

22. Kaul, V. and Sapru, B.L. (1973). 'The Phytosociology and biomass production relations of seven meadow lands in Kashmir', *Vegetative*, Vol. 28, pp. 19-39.

23. Singh, G. and Agrawal, D.P. (1976). 'Radiocarbon evidences for deglaciation in north-western Himalaya, India', *Nature*, Vol. 206, p. 232.

24. Dodia, R., Agrawal, D.P. and Vora, A.B. (1985). 'New Pollen data from the Kashmir Bogs: A Summary', *Climate and Geology of Kashmir; the last 4 million years*, Agrawal, D.P., Kusumgar, S. and Krishnamurthy, R.V. (Eds.), New Delhi; Today and Tomorrow's Printers and Publishers, pp. 101-108, ref. pp. 105-107.

25. See ref. 24. p. 105.

26. Runemark, H. (1956a). 'Studies in *Rhizocarpon*, I : Taxonomy of the yellow species in Europe', *Opera botanica*, Vol. 2, Part 1.

27. Innes, J.L. (1985). 'A Standard *Rhizocarpon* nomenclature for lichenometry', *Boreas*, Vol. 14, Part 1, pp. 25-48.

28. Beschel, R.E. (1961). 'Dating rock surfaces by lichen growth and its application to glaciology and physiography (lichenometry)', *Geology of Arctic-2*, Raasch, G.O. (Ed.), Toronto; University of Toronto Press, pp. 1044-1062.

29. Locke, W.W., Andrews, J.T. and Webber, P.J. (1979). 'A manual for Lichenometry', *British Geomorphological Research Group*, Technical Bulletin, No. 26, pp. 1-47.

30. Armstrong, R.A. (1983). 'Growth curve of the Lichen *Rhizocarpon geographicum*', *New Phytologist*, Vol. 94, pp. 619-622.
31. Luckman, B.H. (1977). 'Lichenometric dating of Holocene moraines at Mount Edith Cavell, Jasper, Alberta', *Canadian Journal of Earth Sciences*, Vol. 14, pp. 1809-1822, ref. pp. 1819-1820.
32. Mathews, J.A. (1975). 'Experiments on the reproductibility and reliability of lichenometric dates, Storbreen gletschervorfeld, Jotunheimen, Norway', *Norsk geografisk Tidsckrift*, Vol. 29, pp. 97-109.
33. Beschel, R.E. (1973). 'Lichens as a measure of the age of recent moraines', Translated by W. Barr, republished in *Arctic and Alpine Research*, Vol. 5, No. 4, pp. 303-309.
34. Carrara, P.E. and Andrews, J.T. (1973). 'Problems and Application of Lichenometry to Geomorphic studies, San Juan Mountains, Colorado', *Arctic and Alpine Research*, Vol. 5, No. 4, pp. 373-384.
35. Kaul, M.N. (1982). 'Fluvial and Glacial Geomorphology of the Wanget basin (Sind River), Kashmir valley', *M. Phil Dissertation*, University of Kashmir, Srinagar, pp. 1-103, ref. pp. 65-68.
36. Benedict, J.B. (1967). 'Recent Glacial History of an Alpine Area in the Colorado Front Range U.S.A.: Establishing a Lichen Growth Curve', *Journal of Glaciology*, Vol. 6, No. 48, pp. 817-831.
37. See ref. 31. p. 1820.
38. Innes, J.L. (1983). 'The Development of Lichenometric dating curves for Highland Scotland', *Transactions of Royal Society of Edinburgh Earth Sciences*, Vol. 74, pp. 23-32.
39. See ref. 26, p. 125.
40. Miller, G.H. (1973). Late Quaternary glacial and climatic history of northern Cumberland Peninsula Baffin Island, Northwest Territories Cananda; *Quaternary Research*, Volume 3, pp. 561-583, ref. pp. 572-74.
41. Haines-Young, R.H. (1983). 'Size variation of *Rhizocarpon* on moraine slopes in Southern Norway', *Arctic and Alpine Research*, Vol. 15, pp. 295-305.
42. See ref. 3, p. 282.
43. See ref. 7, p. 325.
44. Ahmad, N. (1979). 'Morainic deposits in Kashmir Himalayas', *Moraines and Varves: Origin/Genesis/Classification*, Proceedings of INQA Symposium on Genesis and orthwest Territories Lithology of Quaternary deposits, Schluchter, Ch. (Ed.), Rotterdam; A.A. Balkema, pp. 59-63, ref. p. 60.
45. See ref. 24, p. 105.

10

Summary and Conclusions

The book has attempted a discussion of the glacial and fluvial geomorphology of the Liddar valley. It has adopted a combination of empirical and genetic approaches to describe and explain the characteristic landscape features and their ensembles in the Liddar valley. The study includes investigation of the mass balance of the existing glaciers, detailed systematic analysis of different erosional and depositional features of glacial and fluvial landscape and reconstruction of palaeogeographic landscape from fossil flora, pollen analysis and lichenometric and glacial-geomorphological evidences. The major findings, drawn from the present study, are as follows:

1. The study suggests that the Liddar basin was peripherally closed with limited detrital inputs during the pre-Pleistocene times. The Pleistocene uplifts were responsible for the development of large valley glaciers which originated at altitude of 4000 metres or more, terminated at altitude of 2000 metres and occupied the northern parts of the Liddar valley.
2. The distribution of the existing glaciers near the valleyhead areas, where the accumulation of the winter snow is smaller than the ablation, provides an evidence for the retreating of the glaciers. The mass balance investigations of the two of the largest glaciers, Kolahoi (West Liddar) and Shishram (East Liddar), show that the glaciers had achieved greater summer ablation rate than that of the winter accumulation during 1983-84. There was a greater net negative mass balance in Kolahoi glacier (3.15×10^6 m^3 equivalent of water) than the Shishram glacier (2.85×10^6 m^3 equivalent of water). This has resulted in rapid glacier wastage and retreat in the former than in the latter. During the past 120 years the glaciers have retreated nearly 2 kilometres from the snout position of A.D. 1857
3. The field observations of different erosional and depositional features and their investigations reflect that the present

landscape is polygenetic, composed of elements produced by glacial and fluvial processes. The glacial processes were responsible for the deposition of polished, unassorted sediments from the existing glacial snouts to the Ganshibal areas. The fluvial environment was prominent during the deposition of stratified assorted sediments beyond Ganshibal area. On the basis of well preserved lateral moraines, the limits of the major glacial advance have been delineated.

4. The height of the huge, high lateral moraines above the valley floor near Pahalgam and height and extent of striated rock surfaces above the roche moutonnee surface near Shishnag lake reveal that the minimum ice thickness responsible for major glaciation was 250 metres. The striation markings on the valleywall surfaces suggest that substantial ice was generated by the growth of large transfluent glaciers.
5. The glacial erosional landscape is characterised by distinctive erosional features with striations etched on their surface. The higher concentration of erosional features and striation marking on their surface in the middle section of the East and the West Liddar valley are suggestive of excessive erosional intensity in those areas which extended parallel to ice movement.
6. The erosional landscape has resulted from the combined action of plucking and abrasion. The investigations on the silt of the meltwater stream of the Shishram glacier show that abrasion alone has lowered the glacial landscape by 1 metre during past 1000 years.
7. The long profiles of the East and the West Liddar glaciated valleys comprise of three series of treads separated by two steep risers. The steep risers have been formed at those sites where advancing glaciers eroded the narrow valleys at the base of the older wider valleys. The steep risers provide undisputed information regarding the nature and intensity of three glacial advances.
8. The cross valley profiles above the Ganshibal locality suggest pronounced change in slope, marking transition from what appears to be high, broad valley floor to steep-walled gorges containing modern streams.
9. A roche moutonnee provides undisputed information regarding the prolonged and vigorous glacial action. Their distribution at three altitudinal zones of the East and the West Liddar valley

are suggestive of three glacial advances experienced during the Pleistocene Epoch. The striation markings on the top of the Pahalgam moutonnees indicate that the East Liddar glacier was more extensive than the West Liddar glacier.

10. The characteristic clustering of cirques at three altitude zones of the East and the West Liddar valleys document the three former periods of cirque occupation by glaciers. They closely correspond with the three altitudinal zones of roche moutonnees and steep risers of the valleys. The shape and size of cirques vary in four mountain groups of the East and the West Liddar valleys depending upon slope, rock type, distance from ice fence and general wind direction. They are large cylindrical finger type in East Liddar mountain group localised near ice fence which lead to thick accumulation of snow due to favourable wind direction, but are rectangular along the West Liddar mountain group due to thin ice cover and unfavourable wind direction. The dominant direction of cirque floor closely correspond with the wind direction of Srinagar.
11. An examination of the glacial deposited landscape reveals a close relationship between the landscape orientation and ice flow direction. A majority of the moraines have lineated topography and their tills show significant preferred pebble orientation parallel to the morainic crest. The uniformity of large coarse fraction in the tills of different lateral morainic groups is largely attributed to single glacial advance. But increase in the boulder frequency in the tills with increasing altitudes suggest their origin to well ordered sequence of retreat moraines. The analysis of distinctive rock types in the moraines determine the source areas of various glaciers.
12. The distinctive location of end moraines at three altitudinal sites of the East and West Liddar valley provide an evidence of three separate phases of deposition during the third interglacial stage.
13. On the basis of radiometric dates of the bog in Buthpathri locality (125 kilometres away from the study area), the chronological framework of the third interglacial stage for the study area has been established. It indicates that deglaciation was initiated around 17000 years ago at the base of the East and the West Liddar valleys and 8000 years ago around the valleyhead basins.

14. The results of lichenometric analysis on the moraines close to existing glaciers suggest their origin to well ordered sequence of retreat moraines of the Holocene period.
15. The pollen analysis shows overwhelming increase of broad-leaved pollen at three different levels of the bog of the region. They document three alternate phases of warm periods during the third interglacial stage.
16. The stratigraphy of the three terrace levels identified along the Liddar river suggest their origin to fluvial and fluvio-glacial agencies. Their formation date back to the three interglacial stages of the Pleistocene Epoch.
17. The significant feature of the present work has been the palaeogeographic reconstruction of the glacial landscape from the palaeobotanical, lichenometric and glacial-geomorphological evidences. These evidences suggest that the present landscape has resulted from three glacial advances followed by retreats. The first glacial advance had commenced during the early part of the Middle Pleistocene. The third glacial advance was maximum in extent and it resulted in eroding away all the depositional features of the earlier two advances. It was followed by a retreat in three phases; early Ganshibal, middle Ganshibal and late Ganshibal stades. Some of the earlier investigators have documented 4 to 5 glacial advances followed by retreats on the basis of Karewa terraces and moraine levels, though the actual sequence was much smaller.

Bibliography

Ahlman, H.W. (1948). 'Glaciological research on the North Atlantic Coast', *Royal Geographical Society Research Series* 1, pp. 1-83.

Ahmad, N. (1979). 'Morainic deposits in Kashmir Himalayas', *Moraines and Varves : Origin/Genesis/Classification*, Proceeding of an INQA Symposium on *Genesis and Lithology of Quaternary deposits*, Schluchter, Chd. (Ed.), Rotterdum; A.A. Balkama, pp. 59-63.

Ahmad, N. and Hashimi, N.H. (1974). 'Glacial history of Kolahoi Glacier, Kashmir, India', *Journal of Glaciology*, Vol. 13, pp. 279-282.

Ahmad, N. and Mayewski, P.A. (1981). 'Himalayan Glaciation : An Unresolved Issue', *The Geographer*, Vol. 28, No. 2, pp. 21-32.

Andrew, J.T. (1965). 'Surface boulder orientation studies around northwest margin of Barnes Icecap, Baffin Islands, Canada', *Journal of Sedimentary Petrology*, Vol. 35, pp. 735-758.

Andrews, J.T. (1965). 'The Corries of the Northern Nain-Okak section of Labrador', *Geographical Bulletin*, Vol. 7, pp. 129-136.

Andrews, J.T. (1975). *Glacial Systems : An approach to glaciers and their environments*, California; Wadsworth Publishing Company, pp. 1-191.

Andrew, J.T. and Dugdale, R.E. (1971). 'Quaternary history of northern Cumberland Penninsula, Baffin Island, N.W.T. Part V : factors affecting Corrie of glacierisation in Okoa Bay', *Quaternary Research*, Vol. 1, pp. 532-551.

Andrews, J.T. and Smithson, B.B. (1966), 'Till fabrics of the cross-valley moraines of north-central Baffin Island, Northwest territories, Canada, *Bulletin, Geological Society of America*, Vol. 77, pp. 271-290.

Armstrong, R.A. (1983). 'Growth curve of the lichen *Rhizocarpon geographicum*', *New Phytologist*, Vol. 94.

Bakker, J.P. (1965). 'A forgotten factor in the interpretation of glacial stairways', *Zeitschriftt fur Geomorphologie*, Neuefolge 9, pp. 18-34.

Benedict, J.B. (1967). 'Recent Glacial History of an Alpine Area in the Colorado Front Range U.S.A : Establishing a Lichen Growth Curve', *Journal of Glaciology*, Vol. 6, pp. 817-831.

Beschel, R.E. (1961). 'Dating rock surfaces by lichen growth and its application to glaciology and physiography (lichenometry)', *Geology of Arctic-2* Raasch, G.O. (Ed.) Toronto; University of Toronto Press, pp. 1044-1062.

Bhandari, N., and Nijampurkar, V.N. (1981). Glaciers : Our neglected water banks', *Science Today*, No. 12, pp. 60-62.

Bhat, D.K. (1975). 'On the Quaternary geology of Kashmir valley with special reference to stratigraphy and sedimentation', *Geological Survey of India*, Miscellaneous Publication, No. 24, pp. 188-203.

Bhat, D.K. (1978), 'Geological Observation of *margs* on Kashmir valley, India', *Himalayan Geology*, Vol. 8, Part 2, pp. 769-783.

Bion, H.S. and Middlemiss, C.S. (1926). 'The fauna of Agglomeratic Slate Series of Kashmir', *Geological Survey of India*, Vol. 12, pp. 1-55.

Birman, J.H. (1964). 'Glacial geology across the crest of Siera Nevada, California', *Geological Society of America*, Special paper No. 75, pp. 1-80.

Blackwelder, E. (1931). 'Pleistocene glaciation in the Sierra Nevada and Basin ranges', *Bulletin Geological Society of America*, Vol. 42, pp. 865-922.

Boer, G. De (1949). 'Ice-margin features, Lairbreen, Norway', *Journal of Glaciology*, Vol. 1, pp. 332-336.

Bonney, T.G. (1908). 'Fluctuations in Himalayan glaciers', *Nature*, Vol. 77, pp. 201-202.

Boulton, G.S. (1972a). 'The role of thermal regime in glacial sedimentation', *Institute of British Geographers Special Publication*, No. 4, pp. 1-19.

Boulton, G.S. and Vivian, R. (1973). 'Underneath the glaciers', *Geographical Magazine*, Vol. 45, Part 4, pp. 311-319.

Boulton, G.S. (1974). 'Process and patterns of glacial erosion', *Glacial Geology*, Coates, D.R. (Ed.), Binghampton; State University of New York, pp. 41-87.

Butzer, K.W. (1976). *Geomorphology from the Earth*, New York; Harper and Row, pp. 1-463.

Carrara, P.E. and Andrews, J.T. (1973). 'Problems and Application of Lichenometry to Geomorphic studies San Juan Mountains, Colorado', *Arctic and Alpine Research*, Vol. 5, No. 4, pp. 373-384.

Chamberlin, T.C. (1888). 'The rock scourings of great ice invasions', *U.S. Geological Survey 7th Annual Report*, pp. 147-248.

Chapman, C.A. and Rioux, R.L. (1958). 'Statistical study of topography, sheeting and jointing in granite, Acadia National Park, Maine', *American Journal of Sciences*, Vol. 256, pp. 111-127.

Charlesworth, J.K. (1957). *The Quaternary era*, Vol. 1, London; Edward Arnold, pp. 1-900.

Clapperton, C.M. (1971). 'Location and origin of glacial meltwater phenomena in the eastern Cheviot hills', *Proceedings, the Yorkshire Geological Society*, Vol. 38, pp. 361-380.

Clark, J.M. and Lewis, W.V. (1951). 'Rotational movement in cirque and valley glaciers', *Journal of Geology*, Vol. 59, p. 6.

Clayton, K.M. (1965). 'Glacial erosion in the Finger Lakes region (New York State, U.S.A.)', *Zeitschrift fur Geomorphologie*, Band, 9, pp. 50-62.

Conrad, V. and Pollak, L.W. (1950). *Methods in Climatology*, Massachusetts; Harvard University Press, pp. 272-282.

Cotton, C.A. (1941). 'The longitudinal profiles of glaciated valleys', *Journal of Geology*, Vol. 49, pp. 113-128.

Cotton, C.A. (1948). *Landscape : As developed by the processes of normal erosion*, New Zealand; Whitcombe and Tombs Limited, pp. 1-509.

Crosby, I.B. (1945). 'Glacial erosion and the burried Wyoming valley of Pennsylvania', *Bulletin Geological Society of America*, Vol. 56, pp. 389-400.

Dainelli, G. (1922). *Studi Sul glaciale : Spedizione Italiana de Filippi nell Himalaia Caracormume Turchestan, Cinese (1913-1914)*, Ser. II, Bologona; Nicola Zanichelli, pp. 1-658.

Davis, W.M. (1900). 'Glacial erosion in France, Switzerland and Norway', *Proceedings of Boston Society of Natural History*, No. 29, pp. 273-321.

Devendra Pal and Srivastva, V.K. (1982). 'Landform configuration of Karewa floor and its implication on the Quaternary sedimentation, Kashmir Himalaya', *Himalaya : Landform and Processes*, Verma, V.K. and Saklani, P.S. (Eds.), New Delhi; Today and Tomorrow Printers and Publishers, pp. 47-75.

De Terra, H. and Paterson, T.T. (1939). *Studies on the Ice Age in India and Associated Human Cultures*, Washington D.C.; Publication Carneigie Institute, No. 493, pp. 1-345.

Dodia, R. Agarwal, D.P. and Vora, A.B. (1985). 'New Pollen data from the Kashmir Bogs : A Summary', *Climate and Geology of Kashmir; the last 4 million years*, Agarwal, D.P., Kusumgar, S. and Krishnamurthy, R.V. (Eds.), New Delhi; Today and Tomorrow's Printers and Publishers, pp. 101-108.

Embleton, C. (1980). 'Glacial processes', *Processes in Geomorphology*, Embleton, C. and Thornes, J. (Eds.), India; Arnold Heinemann, pp. 272-306.

Embleton, C. and King, C.A.M. (1975). *Glacial Geomorphology*, London; Edward Arnold, pp. 1-573.

Feininger, T. (1971). 'Chemical weathering and glacial erosion of crystalline rocks and the origin of till', *U.S. Geological Survey Professional Paper*, No. 750-C, pp. 65-81.

Flint, R.F. (1971). *Glacial and Quaternary Geology*, New York; John Wiley and Sons, pp. 1-892.

Frakes, L.A. and Crowell, J.C. (1967). 'Facies and palaeogeography of Late Paleozoic diamicite, Falkland Island', *Bulletin Geological Society of America*, Vol. 78, No. 1, pp. 37-58.

Fuchs, Gerhard, (1975). 'Contribution to the geology of NW Himalayas', *Eigentumer Heraugeber and Verleger : Geologisches Bundenstalt*, Wien 32, pp. 1-59.

Fuchs, G. and Gupta, V.J. (1971). 'Palaeozoic Stratigraphy of Kashmir, Kishtwar and Chamba regions', *Verhandlungen der Geologisches Bundenstalt in Wien*, No. 1, pp. 68-97.

Galibert, M.G. (1962). 'Recherches sur les processus d'erosion glaciaires de la Haute Montagne Alpine', *Bulletin Association Geographer Francais*, Vol. 303, No. 4, pp. 8-46.

Gansser, A. (1964). *Geology of the Himalayas*, New York; John Willey and Sons, pp. 1-289.

Ghosh, D.K. (1973). 'Report on Liddar Valley Hydel section', *Geological Survey of India Special paper*, No. 5, pp. 17-21.

Gill W.D. (1952). 'The stratigraphy of Siwalik series in northern Potwar, Punjab Pakistan', *Quarterly Journal of Geological Society*, London, Vol. 107, pp. 375-394.

Glen, J.W., Donner, J.J., and West, R.G. (1957). 'On the mechanism by which stones in till become orientated', *American Journal of Sciences*, Vol. 255, No. 3, pp. 194-205.

Goldthwait, R.P. (1951). 'Development of end moraines in the east-central Baffin Island', *Journal of Geology*, Vol. 59, pp. 567-577.

Goldthwait, R.P. (1966). 'Evidences from Alaskan glaciers of major Climatic Changes', *World climate from 800 to O.B.C.*, Sawyer, J.S. (Ed.), London; Royal Meteorological Society, pp. 40-53.

Graf, W.L. (1970). 'The geomorphology of the glacial valley cross-section', *Arctic and Alpine Research*, Vol. 2, pp. 303-312.

Grinlinton, J.L. (1928). 'Former glaciation of the East Liddar valley, Kashmir', *Memoir, Geological Survey of India*, No. 49, Part 2, pp. 289-388.

Grossal'd, M.G. and Kotlyakov, V.M. (1969). 'Present-day glaciers in the U.S.S.R. and some data on their mass balance', *Journal of Glaciology*, Vol. 8, pp. 9-22.

Gupta, V.J. (1966). 'Palaeontology, Stratigraphy and Structure of the area Southeast of Srinagar', *Unpublished Ph.D. Thesis*, Panjab University.

Gupta, V.J. (1969). 'Palaeozoic stratigraphy of the area southeast of Srinagar', *Research Bulletin, Panjab University*, Vol. 20, Part (I-II), pp. 1-14.

Gupta, V.J. (1970). 'Muth Quartzite Fauna from Kashmir, India', *Research Bulletin*, Panjab University, Vol. 22, Part 1-2, pp. 1-22.

Gupta, V.J. (1971). 'Silurian system of Indian subcontinent', *Research Bulletin of Panjab University*, Vol. 22, Part III-IV, pp. 339-343.

Gupta, V.J. (1973). *Indian Palaeozoic Stratigraphy*, Delhi; Hindustan Publishing Corporation (India), pp. 1-20.

Hack, J.T. (1973a). 'Drainage adjustment in Appalachians', *Fluvial Geomorphology*, Morisawa, M. (ed.) New York; State University, pp. 51-69.

Haines-Young, R.H. (1983). 'Size variation of *Rhizocarpon* on moraine slopes in Southern Norway', *Arctic and Alpine Research*, Vol. 15, pp. 295-305.

Hansen, W.R. (1965). 'The Black Canyon of the Gunnison', *Bulletin, Geological Survey*, No. 1191, pp. 1-76.

Harrison, P.W. (1957a). 'New technique for 3-dimensional fabric analysis of till and englacial debris, containing particle from 3 to 40 millimetres in size', *Journal of Geology*, Vol. 65, No. 1, pp. 98-105.

Harrison, P.W. (1957b). 'A clay till fabric — Its character and origion', *Journal of Geology*, Vol. 65, No. 3, pp. 275-308.

Hashimi, N.H. (1971). 'Notes on the geology of the West Liddar Valley, Kashmir', *Himalayan Geology*, Vol. 1, pp. 284-287.

Hayden, H.H. (1907). 'Notes on certain glaciers in Northwest Kashmir', *Record, Geological Survey of India*, 35, pp. 127-137.

Haynes, V.M. (1968). 'The influence of glacial erosion and rock structure on corries in Scotland', *Geografiska Annaler*, Vol. 50A, pp. 221-234.

Hellaakoski A. (1931). 'On the Transportation of Materials in the Esker of Laitila', *Glacial Deposits*, Goldthwait, R.P. (Ed.), U.S.A.; Benchmark Paper 21, pp. 282-311.

Hoinkes, H.C. and Rudolph, R. (1962). 'Mass Balance studies on Hinterisferner, Otzal Alps, 1951-61', *Journal of Glaciology*, Vol. 4, No. 33, pp. 260-280.

Holland, T.H. (1907). 'A Preliminary Survey of Certain glaciers in N.W. Himalayas-introduction', *Record, Geological Survey of India*, Vol. 35, Part 3, pp. 123-126.

Holmes, C.D. (1941). 'Till fabrics', *Bulletin Geological Society of America*, Vol. 52, No. 9, pp. 1299-1354.

Holmes, C.D. (1944). 'Hypothesis of glacial erosion', *Journal of Geology*, Vol. 52, pp. 184-190.

Holmes, C.D. (1947). 'Kames', *American Journal of Sciences*, Vol. 245, pp. 240-249.

Holmes, G.W. and Moss, J.H. (1955). 'Pleistocene geology of Southwestern Wind River Mountain Wyoming', *Bulletin Geological Society of America*, Vol. 66, pp. 629-654.

Holmssen, G. (1963). 'Glacial deposits of southeastern Norway', *American Journal of Sciences*, Vol. 261, pp. 880-889.

Horman, K. (1974). 'The terraces of Seitikhola : Contribution to Quaternary morphogenesis in central Nepal', *Erdkunda*, Vol. 28, No. 3, pp. 161-176.

Humlum, O. (1981). 'Observations in the basal transport zone of Myrdalsjokull, Iceland', *Annals of Glaciology*, Vol. 2, pp. 71-77.

Innes, J.L. (1983). 'The development of Lichenometric dating curves for Highland Scotland', *Transactions of Royal Society of Edinburgh Earth Sciences*, Vol. 74, pp. 23-32.

Innes, J.L. (1985). 'A Standard *Rhizocarpon* nomenclature for lichenometry', *Boreas*, Vol. 14, Part 1, pp. 25-48.

Jahn, A. (1969). 'Kame terraces in the studies', *Folia Quaternaria*, Vol. 30, pp. 17-20.

Jewtuchowics, S. (1965). 'Description of Eskers and Kames in Gashamnoya and on Bungerbreen, south of Hornsound, Vestspitsbergen', *Journal of Glaciology*, Vol. 5, No. 41, pp. 719-725.

Jhingran, A.G. (1981). 'Geology of the Himalaya', *The Himalaya : Aspect of Change*, Lall, J.S. and Moddie, A.D. (Eds.) Delhi; Oxford University Press, pp. 77-99.

Johnson, D.W. (1904). 'Profile of maturity in Alpine glacial erosion', *Journal of Geology*, Vol. 12, pp. 569-578.

Johnson, D.W. (1944). 'Problems of terrace correlation', *Bulletin Geological Society of America*, Vol. 55, pp. 793-818.

Johnson, W.H. (1976). 'Quaternary Stratigraphy in Illinois : Status current problem', *Quaternary Stratigraphy of North America*, Mahney, W.C. (Ed.), Penn; Stroudsburg, pp. 161-198.

Joshi, R.V., Rajaguru, S.N., Pappu, R.S. and Bopardikar, B.P. (1973). 'Quaternary glaciation and palaeolithic sites in the Liddar Valley (Jammu-Kashmir)', *World Archaeology*, Vol. 5, No. 3, pp. 369-379.

Kachroo, P. (1985). 'A Note on Flora of Kashmir', *Geography of Jammu and Kashmir*, Husain, M. (Ed.), New Delhi; Ariana Publishing House, pp. 32-44.

Kapoor, H.M. (1979). 'Gondwana of Kashmir : A Reappraisal', *Fourth International Gondwana Symposium, Calcutta, India*, 1977, Delhi, Hindustan Publishing Corporation, Volume 2, pp. 443-462.

Kaul, M.N. (1982). 'Fluvial and Glacial Geomorphology of the Wangat basin (Sind River), Kashmir Valley', *M. Phil Dissertation*, University of Kashmir, Srinagar, pp. 1-103.

Kaul, M.N. (1986). 'Mass Balance of Liddar Glaciers', *Trans. Inst. Indian Geographers*, Vol. 8, No. 2, pp. 95-112.

Kaul, M.N. (1988). 'Lichenometric dating of Holocene Moraines in Kashmir Valley', *National Academy of Sciences, India*, 58th Annual Session, pp. 57-58.

Kaul, M.N. (1988) 'Geometry and Mass Balance Eastimates of Kolahoi Glaciers', *National Academy of Sciences, India*, 58th Annual Session, p. 61.

Kaul, V. and Sapru, B.L. (1973). 'The Phytosociology and biomass production relations of seven meadow lands in Kashmir', *Vegetative*, Vol. 28, pp. 19-39.

Khan, A.A. and Srivastava, M.C. (1981). 'An evaluation of textural parameters of Quaternary terraces, Dhauli Ganga, Garhwal,

Himalaya, U.P., *Symposium on three decade development in Petrology, Minerology and Petrochemistry in India.*

King, C.A.M. (1959). 'Geomorphology in Austerdalen, Norway', *Geographical Journal*, Vol. 125, pp. 357-369.

King, C.A.M. (1978). *Techniques in Geomorphology*, London; Edward Arnold, pp. 1-341.

King, C.A.M. (1980). 'Threshold in Glacial Geomorphology', *Thresholds in Geomorphology*, Donald, R. Coates and John, D. Vitek, (Eds.), London; George Allen and Unwin Ltd., pp. 290-297.

King, C.A.M. and Buckley, J. (1968). 'The analysis of the stone size and shape in Arctic environments', *Journal of Sedimentary Petrology*, Vol. 38, pp. 200-214.

Koul, B.L. (1976). 'Structures of Liddar Valley Anticline', *International Seminar on Himalayan Geology*, New Delhi, p. 59.

Krenek, L. and Bhawan, U., (1945). 'Recent and past glaciation of Lahaul', *Indian Geographical Journal*, Vol. 20, Part 3, pp. 93-102.

Krishnan, M.S. (1960). *Geology of India and Burma*, Madras; Higginbothams Pvt. Ltd., pp. 1-603.

Lamplugh, G.W. (1908). 'The gorge and basin of the Zambezi below the Victoia Falls, Rhodesia', *Geographical Journal*, Vol. 31, pp. 133-152, 287-303.

Lee, H.A. (1965). 'Investigation of eskers for mineral exploration', *Geological Survey of Canada*, Vol. 65, No. 14, pp. 1-77.

Lewis, W.V. (1947b). 'Valley steps and glacial valley erosion', *Transaction and Papers, Institute of British Geographers*, No. 17, pp. 19-44.

Lewis, W.V. (1949). 'An esker in process of formation, Boverbreen, Jotunheim, 1947', *Journal of Glaciology*, Vol. 1, pp. 314-319.

Lewis, W.V. (1954). 'Pressure release and glacial erosion', *Journal of Glaciology*, Vol. 2, pp. 417-422.

Lewis, W.V. (1960) 'Norweigian cirque glaciers', *Royal Geographical Society Research Series*, No. 4, pp. 1-104.

Lindsey, D.A. (1969). 'Glacial Sedimentology of Gowganda formation, Ontario', *Bulletin Geological Society of America*, Vol. 86, pp. 1693-1700.

Linton, D.L. (1963). 'The forms of glacial erosion', *Transactions and Papers, Institute of British Geographers*, No. 33, pp. 1-28.

Locke, W.W., Andrews, J.T. and Webber, P.J. (1979). 'A manual for Lichenometry', *British Geomorphological Research Group Technical Bulletin*, No. 26, pp. 1-47.

Luckman, B.H. (1977). 'Lichenometric dating of Holocene moraines at Mount Edith Cavell, Jasper, Alberta', *Canadian Journal of Earth Sciences*, Vol. 14, pp. 1809-1822.

Lydekkar, R. (1883). 'Notes on geology of Kashmir, Kishtwar and Pangi', *Memoir, Geological Survey of India*, Vol. 22, pp. 1-108.

Manley, G. (1959). 'Interpreting the meteorology of the late and post-glacial', *Palaeogeogr. Palaeoclimatol. Palaeoecol.*, Vol. 10, pp. 163-175.

Mannerfelt, C.M. (1949). 'Marginal drainage channels as indicators of the gradients of the Quaternary ice caps', *Geografiska Annaler*, Vol. 31, pp. 194-199.

Marh, B.S. (1986). *Geomorphology of the Ravi River*, New Delhi; Inter-India Publication, pp. 1-86.

Mithal, R.S. (1968). 'The Physiographical and Structural Evolution of the Himalaya', *Mountain and Rivers of India*, New Delhi; 21st International Geographical Congress, pp. 41-81.

Mathews, J.A. (1975). 'Experiments on the reproductibility and reliability of lichenometric dates, Storbreen gletschervorfeld, Jotunheimen, Norway', *Norsk geografisk Tidsskrift*, Vol. 29, pp. 97-109.

Matthes, F.E. (1930). 'Geological history of the Yosemite valley', *U.S. Geological Survey Professional paper*, No. 160, pp. 1-137.

Mayewski, P.A. and Ahmad, N. (1981). 'Reconnaissance glacio-chemical studies in the Indian Himalayas', *Proceedings of the Eastern Snow Conference*, New York, 38th Annual meeting, pp. 45-48.

Mayewski, P.A. and Jeschke, P.A. (1979). 'Himalayan and Trans-Himalayan glacier fluctuations since A.D. 1812', *Arctic and Alpine Research*, Vol. 11, No. 3, pp. 267-287.

Mayewski, P.A. and Pregent, G. (1978) *Results of the 1978 Athabasca glacier expedition*, New Hampshire; Department of Earth Sciences, University of New Hampshire, pp. 1-56.

McCabe, L.H. (1939), 'Nivation and Corries erosion in West Spitsbergen', *Geographical Journal*, Vol. 94, pp. 447-465.

McKenzie, G.D. (1969). 'Observations on Collapsing kame terrace in Glacier Bay National Monument, South-east Alaska', *Journal of Glaciology*, Vol. 8, pp. 413-426.

Medlicott, H.B. (1868). 'The Alps and Himalayas : A Geological Comparison', *Quarterly Journal of Geological Society*, London, Vol. 24, pp. 34-52.

Meher-Homji, V.M. (1971). 'The Climate of Srinagar and its Variability', *Geographical Review of India*, Vol. 33, No. 1, pp. 1-14.

Meier, M.F. (1962). 'Proposed definitions of glacier mass budget term', *Journal of Glaciology*, Vol. 4, pp. 252-261.

Meier, M.F. (1965). 'Glaciers and Climates', *The Quaternary of the United States*, Wrigh, H.E. and Frey, D.G. (Eds.) New Jersey; Princeton University Press, pp. 795-805.

Meier, M.F. and Post, A.S. (1962). 'Recent variations in mass net budgets of glaciers in western North America', *International Association of Scientific Hydrology*, Publication, Vol. 58, pp. 63-77.

Meier, M.F. and Taugborn, W.V. (1965). 'Net budget and flow of South Cascade Glacier, Washington', *Journal of Glaciology*, Vol. 5, pp. 547-566.

Messerli, B., Messerli, P., Pfister, C. and Zumbuhl, H.J. (1978). 'Fluctuations of climate and glaciers in the Bernese Oberland, Switzerland and their geoecological significance, 1600 to 1975', *Arctic and Alpine Research*, Vol. 10, No. 2, pp. 247-260.

Middlemiss, C.S. (1909). 'Gondwana and related marine sedimentary system of India', *Record, Geological Survey of India*, Vol. 37, Part 4, pp. 321-322.

Middlemiss, C.S. (1910). 'A Revision on the Silurian-Triasic sequence in Kashmir', *Record, Geological Survey of India*, Vol. 40, Part 3, pp. 206-260.

Middlemiss, C.S. (1911). 'Section in Pirpanjal Ranges and Sind valley, Kashmir', *Record, Geological Survey of India*, Vol. 41, Part 2, pp. 1-30.

Miller, G.H. (1973). 'Late Quaternary glacial and climatic history of northern Cumberland Peninsula, Baffin Island, North West Territories Canada', *Quaternary Research*, Vol. 3, pp. 561-583.

Mokievsky, Z.O. (1974). 'Analysis of mass balance values and their accuracy for Sentinel glacier, British Columbia Canada', *Inland Water Directorate Resource Branch*, scientific series No. 31.

Moss, J.H. and Ritler, D.F. (1962). 'New evidence regarding the Binghampton substage in the region between the Finger Lake and Catskill, New York', *American Journal of Sciences*, Vol. 260, No. 2, pp. 81-106.

Mukerji, A.B. (1976). 'Choc terraces of the Chandigarh Siwalik hills', *Revue de Geomorphologie Dynamique*, Vol. 25, No. 1, pp. 1-17.

Nakazawa, K., Kapoor, H.M., Ishi, K., Bando, P., Okimura, Y.K. and Tokuoka, T. (1975). 'The Upper Permian and Lower Triassics in Kashmir, India', *Memoir of Faculty of Science, Koyoto University Serial Geology-Minerology*, Vol. 42, pp. 1-106.

Nawani, P.C. and Khan, A.A. (1982). 'Report on Quaternary geology and geomorphology of Chenab basin between Kishtwar and Ramban, J & K', *Geological Survey of India Publication*, No. 12, pp. 1-23.

Neve, E.F. (1910). 'Mt. Kolahoi and its northern glacier', *Alpine Journal*, Vol. 25, pp. 39-42.

Niewiarowski, W. (1963). 'Types of Kames occurring within the area of the last glaciation in Poland compared with kames known from other regions', *Report of 6th Conference of International Association of Quaternary Research*, Warsaw; Lodz, Vol. 3, pp. 475-485.

Norris S.E. and White, G.W. (1961). 'Hydrological significance of buried valleys in glacial drift', *U.S. Geological Survey Professional paper*, 424B, pp. 1-78.

Nye, J.F. (1965). 'A numerical method of inferring the budget history of glacier from its advance and retreat', *Journal of Glaciology*, Vol. 5, pp. 589-607.

Odell, N.E. (1963). 'The Kolahoi Northern Glacier, Kashmir', *Journal of Glaciology*, Vol. 4, pp. 633-635.

Ollier, C.D. (1969). *Weathering*, Edinburgh; Oliver and Boyd, pp. 1-30.

Ostrem, G. (1961). 'A new approach to end moraine Chronology', *Geografiska Annaler*, Vol. 43, pp. 418-419.

Ostrem, G. (1966). 'The height of glacial limit in southern British Columbia, Alberta', *Geografiska Annaler*, Vol. 48A, pp. 126-138.

Ostrem, G. (1966), 'Mass balance studies on glaciers in Western Canada 1965', *Geographical Bulletin*, Vol. 8, pp. 81-107.

Ostrem, G. and Stanely, A.D. (1966). *Glacier mass balance measurements, A manual for fieldwork*, Canada, Department of Mines and Technical Surveys, Ottawa.

Pal, D. and Srivastava, M.C. (1982). 'Landform configuration of the Karewa Floor and its implication on the Quaternary Sedimentation-Kashmir Himalaya', *Himalaya : Landform and Processes*, Verma, V.K. and Saklani, P.S. (Eds.), New Delhi; Today and Tomorrow Printers and Publishers, pp. 47-78.

Pareek, H.S. (1977). 'On the Studies of Agglomeratic Slates and Panjal traps in Jhelum, Liddar and Sind Valley, Kashmir', *Record, Geological Survey of India*, Vol. 107, pp. 12-37.

Pascoe, E.H. (1959). *A Manual of the Geology of India and Burma*, Vol. II, Calcutta; Manager, Government of India Press, pp. 485-1343.

Peterson, W.S.B. (1969). *The Physics of Glaciers*, Canada; Pergamon Press, Toronto, pp. 1-250.

Peterson, W.S.B. and Savage, J.C. (1963). 'Geometry and movement of Athabasca glacier', *Journal of Geophysical Research*, Vol. 68, No. 15, pp. 4513-4520.

Penck, A. (1905). 'Glacial features in the surface of the Alps', *Journal of Geology*, Vol. 13, No. 1, pp. 1-17.

Penck, A. (1973), 'Attempt at a Classification of Climate on Physiographic Basin', *Climate Geomorphology*, Derbyshire, E. (Ed.) London; Macmillan, pp. 51-60.

Peterson, D.N. (1970). 'Glaciological studies on the Casement glacier, southeast Alaska', *Institute of Polar Studies*, Report No. 56, pp. 1-161.

Porter, S.C. (1970). 'Quaternary glacial record in Swat Kohistan, West Pakistan', *Bulletin Geological Society of America*, Vol. 81, pp. 1421-1446.

Price, R.J. (1964). 'Land forms produced by wastage of the Casement glacier, southeastern Alaska', *Institute of Polar Studies*, Report No. 9, pp. 1-164.

Price, R.J. (1969). 'Moraines, sandar, kames and eskers, near Breidamerku-Jokull, Iceland', *Transactions and Papers, Institute of British Geographers*, No. 46, pp. 17-43.

Price, R.J. (1973). *Glacial and fluvioglacial landforms, Edinburgh*; Oliver and Boyd, pp. 1-240.

Puri, G.S. (1957). 'Preliminary observations on Phyto-Geographical changes in the Kashmir valley during Pleistocene', *The Palaeobotanist*, Vol. 6, Part 1, pp. 1-17.

Puri, G.S. (1984). 'A Preliminary notes on the Pleistocene floras of Karewa formation of Kashmir', *Journal of the Geological Mining and Metallurgical Society of India*, Vol. 50, pp. 61-66.

Raina, V.K. (1963). 'A note on some glaciological observations in Garhwal Himalayas', *Indian Minerals*, Vol. 17, Part 2, pp. 159-163.

Raina, V.K. and Kaul, M.K. and Singh, S. (1977). 'Mass balance studies of Gara glacier', *Journal of Glaciology*, Vol. 18, No. 80, pp. 415-423.

Raina, V.K., Bhattacharya, V. and Pattnaik, S. (1973). 'Zamu Glacier', *Record, Geological Survey of India*, Vol. 105, No. 2, pp. 95-106.

Raza, M., Ahmad, A. and Mohamad, A. (1978). *The Valley of Kashmir* : Vol. I, New Delhi; Vikas Publishing House, pp. 1-148.

Robin, G. de Q. (1955), 'Ice movement and temperature distribution in glaciers and ice sheets', *Journal of Glaciology*, Vol. 2, pp. 523-532.

Rothlisberger, H. (1968). 'Erosive processes which are likely to accentuate or reduce the bottom relief of valley glaciers', *International Association of Scientific Hydrology*, Vol. 79, pp. 87-97.

Rudberg, S. (1973). 'Glacial erosion forms of medium size — A discription based on four Swedish case studies', *Zeitschrift fur Geomorphologie*, pp. 33-48.

Runemark, H. (1956a). 'Studies in *Rhizocarpon*, I : Taxonomy of the yellow species in Europe', *Operabotanica*, Vol. 2, Part 1.

Russel, R.J. (1940). 'Quaternary History of Louisiana', *Bulletin, Geological Society of America*, Vol. 51, p. 1214.

Sankalia, H.D. (1971). 'New evidence of Early man in Kashmir', *Current Anthropology*, Vol. 12, No. 45, pp. 558-562.

Sapru, B.L., Dhar, U. and Kachroo P. (1975). 'Vegetation Studies in Jhelum Valley', *Botanique*, Vol. 6, pp. 151-164.

Sastri, C.G.K. (1961). 'Report on Systematic geological mapping in Liddar valley, Anantnag district, Kashmir', *Geological Survey of India*, Misc, Pub. No. 2, pp. 33-39.

Schlagintweit, H., Schlagintweit, A. and Schlagintweit, R. (1861). *Results of Scientific Mission to India and High Asia; with Atlas of panoramas and views*, London; Leipzig, pp. 1-98.

Schytt, V. (1959). 'The glaciers of the Kebnekajse massif', *Geografiska Annaler*, Vol. 41. pp. 213-217.

Schytt, V. (1962). 'Mass Balance Studies in Kebnekajse', *Journal of Glaciology*, Vol. 4, pp. 281-288.

Schytt, V. (1967). 'A Study of Ablation gradient' *Geografiska Annaler*, Vol. 49A No. 2-4, pp. 327-332.

Seddon, B. (1954). 'Late-glacial Cwm glaciers in Wales', *Journal of Glaciology*, Vol. 3, pp. 94-99.

Seeber, L. and Gornitz, V. (1983). 'River profiles along the Himalayan arc as indicators of active tectonics', *Techtonophysics*, Vol. 92, pp. 335-367.

Shah, S.K. (1972). 'Stratigraphic studies of Lower Palaeozoic sequence of Anantnag district, Kashmir', *Himalayan Geology*, Vol. 2, pp. 468-480.

Sharma, V.P. and Sehgal, M.N. (1976). 'Some of the new observations on Stratigraphy and Structure of Liddar Valley anticline, Kashmir', *Geological Survey of India, Kashmir Circle*, No. 14, pp. 56-66.

Sharp, R.P. (1969). 'Semiquantitative Differentiation of Glacial moraines', *Journal of Geology*, Vol. 77, pp. 68-91.

Sharp, R.P. and Birman, J.H. (1963). 'Additions to the classical sequence of Pleistocene glaciations, Sierra Nevada, California', *Bulletin, Geological Society of America*, Vol. 74, pp. 1079-1086.

Shreve, R.L. (1972). 'Movement of water in glaciers', *Journal of Glaciology*, Vol. 11, pp. 205-214.

Singh, D.P. (1974). 'A report on the systematic geological mapping of Liddar Valley, Anantnag district, Jammu and Kashmir', *Geological Survey of India*, No. 34, pp. 47-58.

Singh, G. (1963). 'A Preliminary Survey of Postglacial Vegetational History of Kashmir Valley', *The palaeobotanist*, Vol. 12, No. 1, pp. 73-108.

Singh, G. and Agrawal, D.P. (1976). Radiocarbon evidences for deglaciation in north-western Himalaya, India', *Nature*, Vol. 206, p. 232.

Singh, I.B. (1982). 'Sedimentation pattern in the Karewa basin Kashmir Valley, India and its geological significance', *Journal of Palaeontological Society, India*, Vol. 27, pp. 71-110.

Sissons, J.B. (1958b). 'The deglaciation of part of East Lothian', *Transactions and Papers, Institute of British Geographers*, No. 25, pp. 59-78.

Srikantia, S.V. and Bhargava, O.N. (1983). 'Geology of the Palaeozoic sequence of the Kashmir Tethys Himalayan basin in the Liddar Valley, Jammu and Kashmir', *Geological Society of India*, Vol. 24, No. 7, pp. 363-377.

Stenborg, T. (1969). 'Studies of internal drainage of glaciers', *Geografiska Annaler*, Vol. 51A, pp. 13-41.

Strachey, R. (1848). 'Notes on the motion of glaciers of Pindari in Kumaon', *Journal of the Asiatic Society of Bengal*, Vol. 17, pp. 203-205.

Sugden, D.E. (1968). 'The selectivity of glacial erosion in the Cairngorm mountains, Scotland', *Transactions and Papers, Institute of British Geographers*, No. 45, pp. 79-92.

Sugden, D.E. (1969). 'The age and forms of corries in the Cairngorm mountains, Scotland', *Scotish Geographical Magazine*, Vol. 83, pp. 34-46.

Sugden, D.E. (1970). 'Landforms of deglaciation in the Cairngorm mountains, Scotland', *Transactions and Papers, Institute of British Geographers*, No. 51, pp. 201-219.

Sugden, D.E. and John, B.S. (1976). *Glaciers and Landscape*, London; Edward Arnold Ltd., pp. 1-376.

Suggate, R.P. (1958). 'Late Quaternary deposits of the Christchurch metropolitan area, New Zealand', *New Zealand Journal of Geology and Geophysics*, No. 1, pp. 103-122.

Svensson, H. (1959). 'Is the Cross-section of glacial valley a parabola'? *Journal of Glaciology*, Vol. 3, pp. 362-363.

Tewari, A.P. (1971). 'A short report on glacier studies in Himalayan mountains by the Geological Survey of India', *Stud. Geomorphol. Carpatho-Balcancia*, Vol. 5, pp. 173-181.

Tewari, A.P. (1973). 'Recent changes in the position of the snout of Pindari glacier (Kumaon Himalayas) Almora district, U.P., India', *The Role of Snow and Ice in Hydrology, Proceedings of the Banff Symposia*, Vol. 2, Unesco-WMO-IAHS, pp. 114-1149.

Tewari, A.P. and Jangpangi, B.S. (1962). 'The retreat of Snout of Pindari glacier', in *Variations of the regime of existing glaciers symposium, commission of snow and ice*, U.S.A., Association of Scientific Hydrology Publication, Vol. 58, pp. 245-246.

Thomas, M.F. (1974). *Tropical Geomorphology*, London; Macmillan, pp. 1-332.

Tiku, A.K. Krishnamurthy, A.C. and Gupta, U.P. (1974). 'Report on Liddar Valley Hydel Scheme', *Geological Survey of India*, No. 35, pp. 1-30.

Totten, S.M. (1969). 'Overridden recessional moraines of north central Ohio', *Bulletin, Geological Survey of America*, Vol. 80, pp. 1931-1946.

Trainer, F.W. (1973). 'Formation of Joints of bedrock by moving glacial ice', *U.S. Geological Survey Journal Research*, Vol. 1, Part 2, pp. 229-236.

Trenhaile, A.S. (1975). 'Cirque elevations in the Canadian Cordillera', *Annals of Association of American Geographers*, Vol. 65, No. 4, pp. 518-529.

Tricart, J. (1969). *Geomorphology of Cold Environments*, London; Macmillan, pp. 1-320.

Tricart, J. and Cailleux, A. (1962). *Le modele glaciaire et nival*, Paris; SEDS, pp. 1-508.

Vishnu Mittre (1967). 'Floristic and ecological reconsiderations of the Pleistocene plant impressions from Kashmir', *The Palaeobotanist*, Vol. 16, Part 3 and 4, pp. 308-329.

Vishnu Mittre and Sharma, B.D. (1966). 'Studies of post glacial vegetational history from the Kashmir valley', *The Palaeobotanist*, Vol. 15, No. 1 and 2, pp. 200-204.

Vivian, R. (1970). 'Hydrologie et erosion sous-glaciaires', *Revue, Geog. Alp.*, Vol. 58, Part 2, pp. 241-264.

Vivian, R. (1975). *Les glaciers des Alpes Occidentales*, Grenoble; Allier, pp. 1-513.

Vohra, C.P. (1970). 'A note on the terraces of Sutlej River in Dhar-Slappar area district Bilaspur (H.P.)', *Centre of Advance Study in Geology Publication*, Vol. 7, pp. 119-126.

Vohra, C.P. (1981). 'Himalayan Glaciers', *The Himalaya : Aspect of Change*, Lall, J.S. and Moddie, A.D. (Eds.), Delhi; Oxford University Press, pp. 138-151.

Wadia, D.N. (1934). 'The Cambrian-Triassic sequence of North-Western Kashmir (parts of Muzzafarabad and Baramula Districts)', *Record, Geological Survey of India*, Vol. 68, Part 2, pp. 121-176.

Wadia, D.N. (1937). 'Permo-Carboniferous in the Tertiary Zone of Kashmir Himalayas', *Record, Geological Survey of India*, Vol. 48, Part 2, pp. 1-16.

Wadia, D.N. (1941). 'Pleistocene Ice Age deposits of Kashmir', *Proceedings of the National Institute of Sciences of India*, Vol. 7, pp. 49-60.

Wadia, D.N. (1953). *Geology of India*, London; Macmillan & Co. Ltd., pp. 1-531.

White, G.W. (1968). 'Age and Correlation of Pleistocene deposits at Garfield Heights (Cleveland), Ohio', *Bulletin Geological Society of America*, Vol. 79, pp. 749-752.

Williams, L.D., Barry, R.G. and Andrews, J.T. (1972). 'Application of Computed global rediation to the areas of high relief', *Journal of Applied Meteorology*, Vol. 11, pp. 526-533.

Wisnciwski, E. (1973). 'Genesis of Lammi eskers (South Finland)', *Fennia*, Vol. 122, pp. 1-30.

Wrieght, H.E. (1957). 'Stone orientation in Wadena drumlin field, Minnesota', *Geografiska Annaler*, Vol. 39, pp. 19-31.

Zeuner, E.F. (1959). *The Pleistocene period*, London; Hutchinson, pp. 1-447.

Zumberge, J.H. (1955). 'Glacial erosion in tilted rock layers', *Journal of Geology*, Vol. 63, pp. 149-158.

Zumbuhl, H.J. (1980). *Die Schwankungen der Grindelwaldgletscher : in den historischen Bild-und Schriftquellen des 12. bis 19. Jahrhunderts*, Basel; Birkhauser Verlag, pp. 1-187.

Index